P9-AES-334

THE STEPHEN BECHTEL FUND

IMPRINT IN ECOLOGY AND THE ENVIRONMENT

The Stephen Bechtel Fund has

established this imprint to promote

understanding and conservation of

our natural environment.

The publisher gratefully acknowledges the generous support to this book provided by the Stephen Bechtel Fund

and

the August and Susan Frugé Endowment Fund in California Natural History of the University of California Press Foundation.

Introduction
to Water in California

Introduction
to Water in California

Second Edition

David Carle

for Ken & Elizabeth —
Enjoy !

David Carle

UNIVERSITY OF CALIFORNIA PRESS

University of California Press, one of the most distinguished university presses in the United States, enriches
lives around the world by advancing scholarship in the
humanities, social sciences, and natural sciences. Its
activities are supported by the UC Press Foundation and
by philanthropic contributions from individuals and
institutions. For more information, visit www.ucpress.edu.

University of California Press
Oakland, California

© 2016 by The Regents of the University of California

Library of Congress Cataloging-in-Publication Data

Carle, David, 1950–.
    Introduction to water in California/David Carle—
Second edition.
        p.   cm.
    Includes bibliographical references and index.
    ISBN 978-0-520-28789-1 (cloth : alk. paper).—ISBN
0-520-28789-4 (cloth : alk. paper).—ISBN 978-0-520-
28790-7 (pbk. : alk. paper).—ISBN 0-520-28790-8 (pbk. :
alk. paper).—ISBN 978-0-520-96289-7 (ebook).—ISBN
0-520-96289-3 (ebook)
    1. Water-supply—California.   2. Hydrology—
California.   3. Water-supply—California—
Management.   I. Title.
    TD224.C3C3723   2016
    363.6'109794—dc23   2015015513

Manufactured in China

25 24 23 22 21 20 19 18 17 16
10 9 8 7 6 5 4 3 2 1

The paper used in this publication meets the minimum
requirements of ANSI/NISO Z39.48-1992 (R 2002)
(*Permanence of Paper*).

CONTENTS

Acknowledgments ix

Introduction—Water Web: Connected Californians xi

CHAPTER 1. TAPPING INTO A PLANETARY CYCLE

A Great Water Wheel 1
The Vital Molecule 18
"Normal" Weather: Anything but "Average" 24
Droughts, 26 · Floods, 31

CHAPTER 2. CALIFORNIA WATER LANDSCAPE

Pristine Waterscape 37
Groundwater 50
Hydrologic Regions 54
*North Coast Region, 55 · Sacramento River
Region, 60 · North Lahontan Region, 64 · San Francisco
Bay Region, 67 · San Joaquin River Region, 68 ·
Central Coast Region, 73 · Tulare Lake Region, 74 · South
Lahontan Region, 78 · South Coast Region, 82 ·
Colorado River Region, 85*

CHAPTER 3. THE DISTRIBUTION SYSTEM

Expanding Watersheds 89

The State Water Project 95

The Central Valley Project 108

Colorado River Delivery Systems 115

The Los Angeles Aqueduct 123

The Hetch Hetchy Aqueduct 126

The Mokelumne Aqueduct 131

The North Bay 133

CHAPTER 4. CHALLENGES TO CALIFORNIA
WATER MANAGEMENT

Climate Change and the Water Cycle 141

Extinction Is Forever 147

A Thirsty Garden 159

Asking Too Much of the Colorado River
and the Salton Sea 172

Out of Sight, Out of Control 177
Bad News Beneath Your Feet, 182 · *Fracking, 185*

Can You Drink the Water? 188
The Bottled-Water Phenomenon, 194 ·
Mass Medication, 197 · Giardia, *199* ·
Where Does Your Dog Go?, 201

The Problem Is Us 203

CHAPTER 5. MEETING THE CHALLENGES:
CALIFORNIA'S WATER FUTURE

California Water Law and the Public Trust 206

The Delta, a Tunnel Vision, and a Water Bond   210
The Propostion 1 Water Bond (2014),   220

Recycle and Reuse: Localizing Water   221
Stormwater Capture and Graywater Reuse,   227

Squeezing the Sponge: Conservation   231
*A More Logical Landscape, 236 · Conservation
on the Farm, 238*

Sustainable Groundwater   239
Water in the Bank: Groundwater Storage,   241

The Debate over Dams   245
*Build More Behemoths?, 247 · Off-Stream Dams?, 249 ·
Raise Existing Dams?, 251 · Raze Existing Dams?,   252*

Transfers: Water as a Commodity   257

Clean Water   260

Ecosystem Restoration   262

Lemonade from Lemons: Is Desalination
Viable?   266

Will There Be Enough Water?   272
*Integrated Water Management, 272 · What Future
Do You Choose?,   273*

Acronyms and Abbreviations   279

Historical Timeline   281

Agencies and Organizations   291

References   295

Photo Credits   307

Index   309

Author Biography   326

ACKNOWLEDGMENTS

My thanks go to the late Dorothy Green for suggesting my name, back in 2003, to UC Press when a book on water was to be added to the distinguished list of titles in the California Natural History Guide series. The successful first edition of *Introduction to Water in California* opened a door for my three additional titles in the series, about air, fire, and earth in California. During the following decade, those books were nurtured by editors Doris Kretschmer, Jenny Wapner, Kim Robinson, and, for this new edition, Blake Edgar. The executive director of the Water Education Foundation, Rita Schmidt Sudman (now retired), read the first draft of the water manuscript. The Foundation's field trips and publications, including its Layperson's Guides to every aspect of California water, are excellent sources of information. Frances Spivey Webber, executive director of the Mono Lake Committee back then, now serving on the State Water Resources Control Board, also read the original text, as did Committee cofounder Sally Gaines. Hydrographer Rick Kattelmann, PhD, provided helpful comments and corrections and several excellent photographs.

Sources for illustrations are cited, but I particularly acknowledge the helpfulness of fisheries scientist Tina Swanson and hydrologist Peter Vorster of the Bay Institute; Arya Degenhardt, communications director for the Mono Lake Committee; Rebecca Boyer, with the Department of Water Resources photographic collection; Michele Nielsen, San Bernardino County Museum archivist; Siran Erysian, GIS specialist for the U.S. Bureau of Reclamation; Norma Craig for the Yosemite National Park slide archives; Barbara Beroza and Linda Eade with the Yosemite Library; Richard Harasick of the Los Angeles Department of Water and Power; Joseph Skorupa with U.S. Fish and Wildlife; photographer Frank Balthis; and Jessica Jewell at TreePeople. The staff at the Water Resources Center Archives have been very accommodating.

Many agencies post online information that was helpful in preparing both text and graphics, but I also appreciate the personal responses to my questions from the Coachella Valley Water District, San Bernardino Valley Municipal Water District, Desert Water Agency (in Palm Springs), Mojave Water Agency, Santa Clara Valley Water District, Solano County Water Agency, and the West Basin Water Replenishment District.

Reports published by the Pacific Institute, in Oakland, are invaluable to anyone interested in California water topics. I check, daily, the enormously helpful Maven's Notebook blog for current headline water stories and reports.

With all this expert help, any errors that slipped by are the author's alone.

Janet Carle, my wife, the first reader of all my books, has saved other readers from an overabundance of commas and my occasional slips into murky prose. This edition is, again, dedicated to Janet, who keeps me on track.

Introduction

Water Web: Connected Californians

A California family returned home from a summer outing. Their favorite beach had been posted with closure signs because of contaminated water, but they had found another spot down the coast. Now there was a rush for the bathrooms; toilets flushed several times; then the daughter claimed the first shower. While Mom loaded sandy bathing suits and towels into the washing machine, Dad began rinsing lettuce, tomatoes, and fruit at the kitchen sink. Their son was out in the driveway, energetically hosing salt spray off the family car. Sudsy water ran down the driveway into the sidewalk gutter, eventually falling into a nearby storm drain.

Mom mixed up a pitcher of iced tea and then settled onto a lounge chair beside the swimming pool. Opening bills, she read aloud to Dad (as he put hamburgers on the grill) from a water company insert titled "The Water We Use Each Day."

"It took eight gallons of water to grow one of those tomatoes you just sliced," she told her husband. "One gallon for each almond nut! That burger patty you're holding took 616 gallons and my cotton jeans...at least 1,800 gallons of water."

Dad started a sprinkler going on the lawn and noticed that the swimming pool level was down. The weather was warm and dry, and the kids' pool party had splashed plenty of water out the day before.

"Landscaping consumes about half the water Californians use at home," Mom added, still reading. "Toilets use 20 percent and showers another 18 percent."

"Our water meter must be spinning like crazy right now," Dad said, wondering when their daughter would emerge from the shower.

About six weeks earlier, a snow patch had finished melting near the summit of a Sierra Nevada mountain peak. Liquid and flowing again after five months in cold storage, the water soaked into the ground and began to percolate downhill, pulled by gravity. Nearby tree roots absorbed much of it, but the rest eventually seeped out into a small creek at the base of the hill. The sun and wind evaporated a bit of the water. Animals drank a little. Some passed through a trout's mouth and gills, losing dissolved oxygen and carrying off a bit of carbon dioxide.

For millennia, water that traveled along this particular part of the water cycle had cascaded down the steep eastern face of the Sierra Nevada, carving a canyon as it went, and finally entered a salty inland sea called Mono Lake. Algae, brine shrimp, and millions of birds took advantage of that oasis in the desert. From there water had nowhere to go but up, reentering the atmosphere by evaporation, to someday fall again as rain or snow. But now, much of this snowmelt was diverted into an aqueduct. It began a 350-mile trip southward, finally reaching a storage reservoir in Los Angeles.

In Southern California, it was mixed with other water that had followed even longer routes. Some had originated in snowfields on the western slope of the Sierra Nevada. As that water approached San Francisco Bay via the Sacramento River and Bay-Delta, it was diverted southward into the California Aqueduct. Saltier water traveled up to 1,400 miles from the Colorado River watershed in the Rocky Mountains. Water pumped out of the ground from a local Southern California aquifer joined the mix. That groundwater carried industrial contaminants, but at levels deemed acceptable when diluted.

Now, the water that had entered the Los Angeles Aqueduct system a month earlier was inside a pipe, poised before the water meter of a Southern California home. Each time there was a surge of movement, the meter measured the flow and water moved off through pipes toward the bathrooms, kitchen, laundry room, and yard spigots.

The water from the Eastern Sierra snowfield took its turn sliding past the meter, paused, crept forward, and then began to ooze into the open. A single drop slowly gathered weight at the mouth of the kitchen faucet. At last it fell, straight into the drain. From there, it headed down toward the city's sewage treatment plant. Soon it would return to the sea.

That faucet continued steadily dripping, as it had been doing for many weeks. Another six gallons of water, laboriously harvested from distant environments, dripped down the drain by the end of that day.

An intricate system of dams, aqueducts, and pipes delivers water to people in California. Though water is the essential molecule supporting life on Earth, it can be taken for granted so long as the distribution system quietly works behind the scenes and the California climate cooperates with "normal" weather. Yet the movement of water across the landscape to serve human needs has consequences where water originates and where it emerges from faucets. A full appreciation of today's thoroughly "plumbed" California should foster understanding of the consequences that individual behaviors or community decisions bring for all who share California's water.

Consider how difficult it must be for someone in Los Angeles, at the far end of "the pipe," to realize the connection between a few gallons of water thoughtlessly wasted—or carefully conserved—each day and populations of birds on a salt lake 350 miles away. An environmental battle was fought in the courts and in the arena of public opinion for 16 years over damage caused to Mono Lake, an inland sea east of Yosemite National Park, by stream diversions to Los Angeles. After visiting the lake and seeing it teeming with migratory birds, many tourists are amazed that anyone in California ever opposed complete protection for such a national treasure. Yet the distant, unseen impacts are hard to perceive for millions of urban water users making individual daily choices.

When a new subdivision is built in an urban area, do planners and developers or the families that move into those houses appreciate the connections being made? Water must reach each household through a network of dams, aqueducts, water treatment plants, and delivery pipes. Some water may travel nearly 700 miles from the upper watershed of the Feather River in Northern California. Another branch of the water system extends 1,400 miles up the Colorado River to its headwaters in the Rocky Mountains of Colorado and Wyoming, so that the winter weather of those distant states has become more significant to many Californians' annual water supply than local rainfall.

Federal, state, and local agencies operate systems that tie Californians together through their web of water pipes. Water availability shaped the state's urban growth and development. Extensive aqueduct systems enabled population increases far beyond what regional water limits would have allowed. Those large populations generated wastes and industrial pollutants that have moved into groundwater basins, rivers, and lakes, and diminished the water quality. Looking for more water to serve demands, urban water interests today covet supplies committed to California agriculture. The state's farms feed much of the nation, but they face pressure and financial incentives to market or transfer their water to domestic uses.

The environment remains the ultimate source for water. California has a tremendous range of climates and hydrologic conditions. Snow on the Sierra Nevada becomes the state's largest "reservoir," refilled each winter, then gradually emptying into streams, rivers, and groundwater basins. Every river canyon in California that is suitable for a major dam has been developed. Flowing rivers have been transformed into reservoirs, much of the water now traveling through irrigation ditches and city pipes, leaving some river channels dry. Plants and animals that are just as dependent on water must settle for whatever people are willing to share with them. The transformation of the state's water-dependent habitats is greatly responsible for the length of California's list of endangered and threatened species.

The first chapter in this book considers the water cycle that moves vital molecules of H_2O across the California landscape, creating precipitation and climate patterns that shape the state's relationship with water. Next, the original waterscape of California is compared to today's scene, across 10 distinct hydrologic regions. In the third chapter, the distribution system that carries water between regions to serve human purposes is detailed. Many of the changes in the California waterscape followed the creation of the transport systems described here. The wide range of challenges that have resulted are described in the fourth chapter, as well as a survey of environmental concerns, lost habitat, endangered species, pollution, domestic water quality issues, and of ties between water and growth. Finally, attention turns to ways of addressing these challenges to shape California's water future. Statewide planning, conservation, recycling, groundwater pumping, water marketing, and the future of dams in this state are some of the topics.

This natural history book recognizes the overwhelming role of humanity in the story of California water, to provide a contemporary understanding of the natural waterscape and watersheds of the state and of the extended watersheds that people created by redirecting water across the West. The goal is to help Californians better appreciate the water that emerges from their taps, what it takes to move it there, and what changes occur in environments along the way.

It has been more than 12 years since the manuscript for the first edition of *Introduction to Water in California* was completed, and, in that time, considerable changes and turmoil have stirred the world of California water. The effects of climate change on California's water systems became clearer and increasingly challenging. A 14-year drought on the Colorado River watershed complicated the state's plans for living within its actual allotments from that over-allocated source. Extreme drought in California from 2012 into 2015 brought the driest years on record, and public awareness of those conditions helped win passage of a $7.5 billion bond to fund conservation, recycling, and new storage. California had been the only state in the nation without groundwater management regulation until 2014, when the Sustainable Groundwater Act passed in the legislature, but implementation of new plans by local agencies does not have to be accomplished until 2040 (almost three decades is better than never). The ecosystem of the Sacramento–San Joaquin Delta came close to collapse, triggering court-ordered reductions in pumping that impacted State Water Project (SWP) and Central Valley Project (CVP) customers and fostered a series of new laws, agencies, and projects, including controversial "twin tunnels" to run beneath the Delta as part of the California Water Fix plan. The effects on water quality and supply from fracking, a technique for stimulating oil and natural gas fields to increase yield, became a new water quality and water supply concern. The "State Water Plan, Update 2013," was released in December 2014 to guide water decisions affecting the future. In 2015, the state entered a fourth year of extreme drought. When the annual snow survey on April 1 found a record low five percent of normal snowpack in the Sierra Nevada, Governor Jerry Brown issued the first ever mandatory water conservation directive in California history, requiring a 25 percent cut in urban water use.

Today, overall demand exceeds the supply of developed water in California. Understanding the natural role of water in this environment and the complexities of the interconnected water system is the first step toward wise choices by society's decision makers and by every Californian.

CHAPTER ONE

Tapping into a Planetary Cycle

See how every raindrop and snowflake, every skyborne molecule of H_2O that falls...is also a child of Ocean and Sun.... See how those streams and rivers, as Aldo Leopold pointed out, are "round," running past our feet and out to sea, then rising up in great tapestries of gravity-defying vapor to blow and flow back over us in oceans of cloud, fall once more upon the slopes as rain and snow, then congeal and start seaward, forming the perpetual prayer wheels we call watersheds.

—David Duncan, *My Story as Told by Water*

Especially as I drink the last of my water, I believe that we are subjects of the planet's hydrologic process, too proud to write ourselves into textbooks along with clouds, rivers, and morning dew.

—Craig Childs, *The Secret Knowledge of Water*

A GREAT WATER WHEEL

A partnership between land and a planetary water cycle produces the California climate and shapes the natural landscape of the state. California's weather is generated primarily by westerly winds off the Pacific Ocean. In the winter, low pressure in the northern Pacific sends cold, wet storms to the state. California receives 75 percent of its annual precipitation between November and March, the majority from December through February. The dry weather of summer is associated with a high-pressure "dome" over the Pacific. Such "Mediterranean" climates, with wet winters and summer droughts, occur on the west coasts of continents in the middle latitudes due to global patterns of atmospheric pressure circulating over the oceans.

California's rainfall is heaviest in the north and decreases toward the south (map 1). Eureka, surrounded by redwood rain forests, usually receives more than 50 inches of rain each winter. That North Coast town has as much claim to a "California climate" as Los Angeles does, with only 15 inches on average. It *does* rain in Southern California, contrary to the myth popularized by real estate promoters and Hollywood, and Los Angeles does experience seasons. Winter rains activate the southern California growing season, as dormant plants awake and seeds of annual plants germinate. Summer brings a seasonal drought, and the autumn transition includes hot, dry Santa Ana winds and wildfires. Mountain communities such as Lake Tahoe and Mammoth Lakes experience yet another version of California weather, with six months of winter snow and the brief summer growing season characteristic of alpine landscapes (fig. 1).

California's diverse landscape is responsible for this wide range of precipitation patterns. The state's coastline stretches 800 miles from Oregon to Mexico. A map of California, superimposed over the east coast of the United States, would extend from southern Maine all the way to South Carolina, crossing more than nine degrees of latitude. But California has more diverse weather and climate than the East, because its 100 million acres contain the tallest mountain ranges in the 48 contiguous states and desert basins that lie hundreds of feet below sea level (map 2).

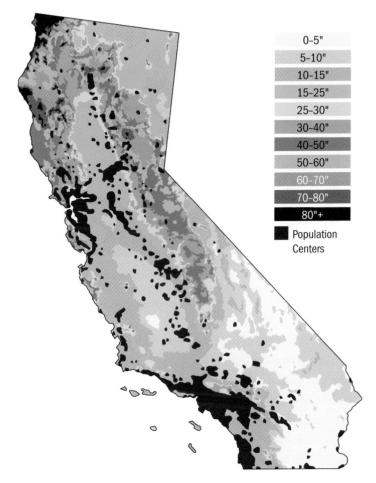

	0-5"
	5-10"
	10-15"
	15-25"
	25-30"
	30-40"
	40-50"
	50-60"
	60-70"
	70-80"
	80"+
	Population Centers

Map 1. Average annual precipitation, in inches, in California (redrawn from Hundley Jr. 2001).

Figure 1. Mammoth Creek. Six months of winter snow and a brief summer growing season are one version of the California climate.

Rainfall and snowfall result when humid air masses blow in from the ocean and interact with the state's mountain ranges. Moist air, moved inland by the prevailing westerlies, pushes up against California's mountain backbones, which wring vapor out of air as it rises, cools, and condenses (fig. 2). Precipitation generally increases two to four inches for each 300-foot rise. Seasonal snowfall totals about two feet at the 3,000-foot elevation in the Sierra Nevada foothills, but increases to 34 feet on Donner Summit, the famous 7,000-foot pass where the Donner party spent a tragic winter. The Sierra Nevada occupies one-fifth of the land area of California and has a major influence on the climate, weather, and water supply of much of the state. Its crest extends 430 miles; 8,000-foot summits in the north rise to over 14,000 feet in the south, intercepting the westerly jet stream at higher and higher elevations. Most of the precipitation in the Sierra

Map 2. Landform provinces in California (redrawn from Schoenherr 1992).

Figure 2. Air cooling and condensing as it rises over mountains.

Nevada falls as winter snow (fig. 3). In Plumas County, north of Lake Tahoe, an average of 90 inches of precipitation falls at 5,000 feet. The same elevation in the southern Sierra receives as little as 30 inches.

As air descends the east side of California's mountain ranges, the process is reversed. Air becomes warmer and holds more of its water vapor. Relatively dry "rain shadows" are the result. The Sierra Nevada rain shadow creates the Great Basin desert. The Coast Ranges produce a rain-shadow effect for the Central Valley too, although a major gap at San Francisco Bay lets more moisture directly strike the northern Sierra Nevada. The Mojave and Colorado Deserts lie in the rain shadow of the southern Sierra Nevada but are primarily influenced by the Transverse and Peninsular Ranges. The Mojave Desert town of Barstow averages only four inches of rain per year; Imperial, farther south in the Colorado Desert, is even drier (fig. 4).

Figure 3. Sierra Nevada precipitation builds the winter snow pack.

Figure 4. The Mojave Desert in the rain shadow of the Sierra Nevada.

A broad cross section through the state, beginning near San Luis Obispo and extending roughly northeastward, intersecting the mountain ranges at right angles, would pass through the Central Valley near Visalia, cross Sequoia National Park, and take in the Owens Valley town of Independence. San Luis Obispo, at the base of low mountains in the Coast Ranges, averages 22 inches of rain; Coalinga, in the Coast Ranges' rain shadow and down on the floor of the Central Valley, receives only seven inches. Farther east, just below the Sierra Nevada foothills, Visalia picks up 11 inches. Giant Forest, in Sequoia National Park, is at 7,000 feet; snow and rain there total 46 inches of precipitation (fig. 5). Independence is in a desert created by the Sierra's rain shadow and averages only five inches of rain. East of the White Mountains, in Death Valley, 178 feet below sea level, annual precipitation is a mere two inches at Greenland Ranch (fig. 6).

California receives almost 200 million acre-feet (MAF) of precipitation in an average year. One acre-foot (AF) equals 325,851 gallons, which would cover a football field one foot deep. Planners commonly figure that an AF serves the annual domestic needs of one to two families, or five to eight people, depending on how wisely it is used and conserved. Water that falls on the state may evaporate back into the atmosphere, be used by plants that then return vapor to the air, or soak deep into groundwater basins. What remains is about 71 MAF of "runoff" water, which moves across the landscape and is the water most accessible to people. Streams draining the sodden North Coast contain about 40 percent of this runoff. The Sacramento River basin generates another 31 percent, mostly originating with the Sierra Nevada snowpack. Snowmelt from the southern Sierra drains into the San Joaquin and Tulare Valleys, producing much of the balance (map 3). The Colorado River receives almost no runoff

Figure 5. Snow at 7,000 feet in Sequoia National Park.

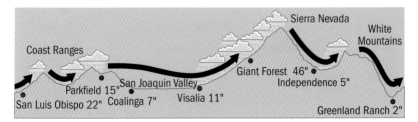

Figure 6. Influence of topography on precipitation; a southwest–northeast cross section of California (redrawn from Durrenberger 1968).

------------------ 100,000–500,000 Acre-feet

◢ Over 500,000 Acre-feet

Map 3. Average annual streamflow in California (redrawn from Durren-berger 1968).

originating inside California, but because the river serves as the state's southeastern border, California receives 4.4 MAF from it. This apportionment, along with Klamath River water out of Oregon, allows water planners to figure on a statewide supply of 78 MAF of annual runoff.

The Sierra Nevada snowpack has historically peaked by April 1 and then begun melting. By midsummer it is gone, except for a few small glaciers and snowfields on north-facing exposures that are shaded from direct sunlight. The delayed release of snowpack water overlaps only partly with the optimum growing season for plants in California. Moisture is most available in the winter, when temperatures are low, and is scarce during the long, warm days that optimize growth. Urban and agricultural water demands are out of sync with the natural runoff pattern, peaking during summer and at their low point during winter. California's natural vegetation evolved adaptations to the local patterns. Many annual plants flower quickly in spring and produce seeds that sleep through the long drought of summer and early autumn. Winter rains break that dormancy. Some perennial shrubs and trees rely on deep root systems to tap water even during the long seasonal droughts. Others go dormant, simply shutting down their metabolisms. Riparian vegetation found along riverbanks and in wetlands benefits from year-round water availability. The New England pattern of four seasons—lush, green springs; hot, wet summers that encourage plant growth; autumn color before leaves are dropped; and freezing winter weather—is found in California only in mountain and foothill river canyons. There plants can keep their roots wet and local hydrologic conditions mimic the New England pattern (fig. 7).

Water moving within California is part of a greater planetary water cycle that includes many circular movements, wheels

Figure 7. Plants watered year-round by a mountain creek.

within wheels (fig. 8). Water is continuously shifting among three "reservoirs": the ocean, the atmosphere, and the land. These are connected by precipitation, evaporation, and plant absorption and transpiration (evaporation through leaf pores). Water is perpetually changing form and traveling the globe. It has been said that we drink the same water the dinosaurs drank. That is not accurate for specific water molecules. During photosynthesis, for

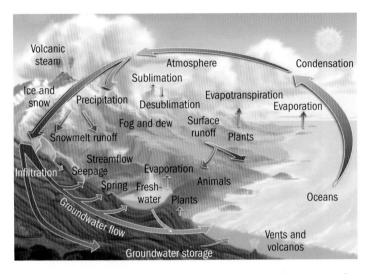

Figure 8. The water cycle: wheels within wheels. From U.S. Department of the Interior, U.S. Geological Survey (John Evans, Howard Perlman). http://ga .water.usgs.gov/edu/watercycle.html.

example, these molecules split into oxygen and hydrogen atoms. Yet it is true that no water is lost in the overall planetary balance; water returns. The respiration of plants and animals recycles it, reversing the photosynthesis equation by consuming oxygen while breaking complex molecules into water and carbon dioxide. Fire, an important decomposition agent in the natural California landscape, produces the same chemical results. And when organisms die and decompose, water is reconstituted.

This planetary recycling is powered by the sun, which evaporates water from the ocean and the land. In photosynthesis, the sun's energy is also what splits the bonds holding water molecules together. Of the water vapor returned to the atmosphere, 16 percent comes from transpiration by land plants (fig. 9); most of the rest comes from the ocean. At any given moment, only a

Figure 9. Evapotranspiration returning water to the
atmosphere.

thousandth of one percent (0.00001) of the planet's total water is
in the air. Yet that small percentage produces thick coastal fogs,
dramatic thunderheads, and drenching downpours. In a journal
entry written during a January storm, John Muir marveled "that
so much rain can be stored in the sky" ([1938] 1979, 335). The
recycling that replenishes atmospheric vapor is so constant and

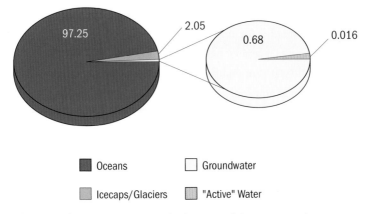

Figure 10. Planetary water reservoirs (percent of planetary water).

voluminous that this water is completely replaced every eight days and the equivalent of all the oceans' water passes through the atmosphere every 3,100 years.

Two-thirds of the Earth's surface is covered by liquid water. Philip Ball wrote, in *Life's Matrix: A Biography of Water* (2001, 22), "We call our home Earth – but Water would be more apt." Over 97 percent is salt water, though, and over two-thirds of the freshwater is locked up in ice caps and glaciers. Less than one percent of the total is available freshwater, with most of that below ground, in aquifers that are never fully accessible. On this watery planet, just 0.016 percent (0.00016) of the precious fluid is "active" freshwater, moving through lakes, rivers, the atmosphere, and living creatures (fig. 10).

The cogs in the water-recycling wheel revolve at different speeds, like different-sized gears meshing inside an enormously complex clock. Vapor evaporated from the surface of the sea may circulate for only a few hours or for days. Deep ocean water may take thousands of years to complete a circuit of evaporation, condensation, and return. Some of the water in the polar ice caps may remain solid for millions of years. The ice in some small Sierra Nevada glaciers has been there for nearly a thousand years. Under certain conditions, groundwater can be trapped in deep, confined aquifers, held back from the water cycle for thousands of years. At its own speed, however, groundwater does participate in the cycle. It feeds springs, rivers, or lakes, and it is replenished when surface water percolates into the ground.

Water may travel for weeks through California's river arteries before finally returning to the sea or terminating its journey in inland waters such as Mono Lake. Almost anywhere along these routes it may be shunted aside, pulled in by the roots of a plant, or drunk by an animal.

Water is essential for life on Earth and is the critical habitat factor that shapes California's ecosystems. As the leaf and root designs of plants adapt to climate, elevation, soil, and topography, both the gathering and the conservation of water are of supreme importance. Bands of different flowers lining a vernal pool sort themselves out by their particular relationships with water. The spiny leaves of a Joshua tree *(Yucca brevifolia)*, like the extremely efficient kidneys of a kangaroo rat, are water-conservation adaptations to life in the desert (fig. 11). Indeed, everywhere in the state—in the wetland marshes rimming San Francisco Bay, the grassy prairies of the Central Valley, the north coast rain forests, the chaparral shrublands of southern California, the foothill oak woodlands, and the pine forests of the Sierra Nevada—all forms of life accommodate to the local availability of water. Photosynthesis requires water, often in enormous amounts. Plants combine water with carbon dioxide to manufacture food for themselves and the herbivores that feed on them; in the process, they replenish the atmosphere with oxygen gas.

Figure 11. Needles and leaves designed to conserve water.

Various mechanisms and behaviors foster "best management practices" for water conservation by living things. At the boundaries between multicellular bodies and the rest of the world, barriers of skin, bark, scales, or mucous membranes regulate water passage in and out. Every living cell has a membrane that encloses and regulates its internal concoction of water and essential chemicals. Multicellular organisms bathe their cells in watery environments. Water management is critical to homeostasis, the maintenance of the internal conditions necessary for life.

We are bodies of water. Humans can live without food for weeks, but die within days when deprived of water. Our bodies are 65 percent water (our brains more than 95 percent); a 150-pound human body contains over 12 gallons of water. We need to replenish about two and a half quarts a day, one-third from drinking and the rest in foods, as we lose water in breath, sweat, and urine. Water is the primary medium for biochemical reactions and a participant in many of the essential processes of life. It helps break down our food, then carries the digestion products to our cells. It regulates temperature and transports dissolved oxygen and carbon dioxide through our circulatory systems. Proteins that rely partly on their shapes to fulfill their jobs as enzymes are folded into those shapes by bonds with water in the fluid of our cells. As cellular metabolism generates wastes, water dissolves them and moves them across filtration membranes in our kidneys, returning them (and the water itself) to the environment.

Water is so essential to us that it is amazing we ever take it for granted. If it is our most precious resource, that is not simply because the supply sometimes grows scarce. H_2O is the vital essence of life on Earth, an almost magical molecule. A full appreciation of our relationship with California water begins at the molecular level.

THE VITAL MOLECULE

Water is so familiar that we seldom give any thought to what sets this particular molecule apart from other substances commonly found in our lives. Unusual characteristics are behind water's critical importance. "Water is life's true and unique medium," Philip Ball has written. "That the only solvent with the refinement needed for nature's most intimate machinations happens to be the one that covers two-thirds of our planet is surely something to take away and marvel at" (2001, 268).

Most solids, liquids, or gases that we encounter naturally are found in just one phase. Minerals, such as silica or calcium carbonate, that form rocks and soils remain solid (unless heated to extremes by volcanic action or movement of the plates that form the Earth's crust). Other elements and compounds, too, stay in a single phase under normal circumstances. "Silicon vapor" is not part of our daily experience or vocabulary. Neither is "liquid wood" or, for that matter, "solid air." Decomposition or digestion breaks molecules apart to build something new, but this is not simply a matter of phase changes. The water molecule, however, is widely abundant on this planet in all three phases: as solid ice, liquid water, and gaseous water vapor (fig. 12). When most other molecules *are* transformed, those changes regularly involve water because it is so nearly ubiquitous, dissolves most anything, and is good at carrying other materials along with it.

The explanation for water's unusual phase character also helps explain why water contains "an invisible flame...that creates not heat but life," as described by David Duncan in *My Story as Told by Water* (2001, 190). Water is a "community molecule." That is, water molecules constantly form, break, and re-form bonds with one another. Those bonds produce a cohesive tendency that is behind most of water's special attributes. Working together, H_2O molecules pick up the colors of the sky, create the pleasing sounds of water and gravity working together, and shape our most beautiful landscapes.

The cohesiveness is explained by the relationship between two hydrogen atoms and one oxygen atom. H-O-H is a polar

Figure 12. Ice crystals, one widely abundant form of water.

molecule, more positively charged at the hydrogen ends and more negatively charged toward the oxygen atom. Because opposite charges attract, hydrogen atoms within a water molecule will orient toward the oxygen in another, nearby. Each H_2O molecule can form such "hydrogen bonds" with up to four others (fig. 13).

This bonding explains many unique properties. H_2O's polarity makes it the "universal solvent" because its charged ends

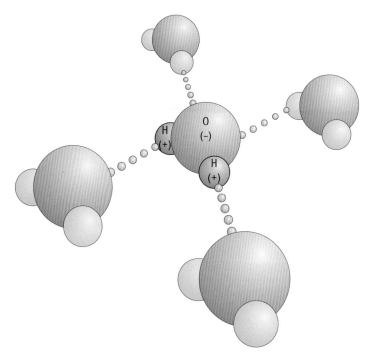

Figure 13. Hydrogen bonds between water molecules.

seek out opposite charges on many other kinds of molecules as well. Water weasels its way between such molecules, separating them and carrying them into solution. It dissolves so many things that truly pure water is rarely, if ever, found in nature. The chemistry of your local water supply reflects whatever rock and soil it has touched. Even atmospheric vapor, distilled to purity by evaporation, soon finds substances to react with in the air. Acid rain is one familiar and unpleasant result.

Bonded cohesiveness also accounts for the ability of California's 300-foot-tall redwood trees to lift water from the ground up to their topmost branches inside cellular tubes—water ropes

Figure 14. A water strider on the surface of a pond, skating on the surface tension created by water molecules.

that move against gravity because tension is applied from the top, where leaves pass water out to the atmosphere. It also explains the surface tension that allows a water strider to skate across the surface of a pond (fig. 14).

Unlike almost every other liquid naturally found on Earth, water expands when it freezes. The expansion begins, oddly, *before* the freezing point is reached. As cooling water drops below 40 degrees F, it suddenly becomes less dense—the opposite of its behavior above that temperature. It takes on a regular crystalline shape with empty space between the oppositely charged hydrogen and oxygen portions of neighboring molecules. Because of this strange behavior, solid water—ice—is less dense than liquid water, and ice floats (fig. 15). If the water in mountain lakes behaved like most liquids, cooling at the surface would cause denser ice to settle to the bottom and, gradually, the lake would freeze solid right up to the surface. Because ice floats, fish

Figure 15. Ice floating on a mountain lake.

and other aquatic creatures can carry on through the winter beneath insulating ice layers that eventually arrest further cooling of the depths. Because water expands when it freezes, mountain residents have to protect pipes from bursting in the winter. The internal liquid environment in all living things must also be protected from freeze-expansion that could destroy cellular membranes and tissues. Freezing water even shapes the California landscape wherever water in cracks expands as ice, causing rock to peel and break.

The cohesiveness of hydrogen bonds means that it requires an unusual amount of energy for water to change phases. Water has high melting and boiling points because bonded molecules resist being pulled apart. Thus, for example, when sweat evaporates, a great deal of heat is carried off, efficiently cooling our bodies. Water also resists too-rapid heating. It takes more heat to raise the temperature of water than to raise that of most other liquid or solid substances by the same amount. Though a sandy beach in the sun gets very hot, the nearby seawater or a lawn bordering the beach remains cool. Both the ocean and the grass heat up slowly and are constantly losing energy through evaporation.

The cohesive attraction between molecules of water also means that there is a direct connection between you and your watershed, through hundreds of miles of pipes, treatment plants, aqueducts, reservoirs, and rivers. Continuous "ropes" of water may extend from a San Diego faucet all the way to the northern Sierra Nevada and the Colorado Rockies. "Pull" from your end and water molecules transmit that tiny force, reacting all the way up the line.

"NORMAL" WEATHER: ANYTHING BUT "AVERAGE"

And it never failed that during the dry years the people forgot about the rich years, and during the wet years they lost all memory of the dry years. It was always that way.
—John Steinbeck, *East of Eden*

The normal climate of California includes droughts and years that are particularly wet (fig. 16). Very rarely does California weather actually match long-term averages. The state's average annual runoff totals 71 MAF, but the range is tremendous—as little as 15 MAF during the severe drought year of 1977, but up to 135 MAF in the exceptionally wet winter of 1983. The annual volume in Sierra Nevada rivers can be 20 times as great in very wet years as in very dry years (fig. 17).

Such year-to-year variations are partially tied to variable Pacific Ocean temperatures known as the El Niño Southern Oscillation. When warmer ocean currents shift eastward in the Pacific, toward the coasts of North and South America, jet streams

Figure 16. A sign of drought, part of the normal weather
cycle in California.

and storm tracks overhead shift accordingly. Results vary, but
strong El Niño winters often lead to wetter than normal winters
for northern California, with a smaller effect in the southern part
of the state. El Niño events are interspersed with La Niña events,
with colder-than-normal ocean temperatures and, usually, below-
normal precipitation for California. This cycling between El
Niño and La Niña, every three to seven years, is a natural phenomenon that can be traced back thousands of years (figs. 18a, 18b).

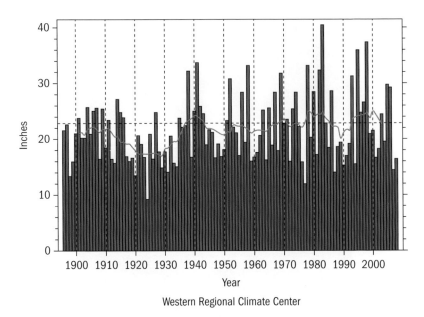

Figure 17. California statewide precipitation variability, 1896 to 2009 water years, October–September (redrawn from California Department of Water Resources 2013b).

NOTE: Orange line denotes 11-year running mean.

The Pacific Decadal Oscillation (PDO) is a separate 10- to 20-year ocean-atmosphere cycle producing less spectacular effects than El Niño, but with a longer duration. During positive phases of the PDO, an El Niño event can be amplified, while negative phases of the decades-long cycle may accentuate La Niña effects and lead to droughts.

Droughts

No simple criteria define a drought. Water providers in California may announce a drought emergency whenever there is too

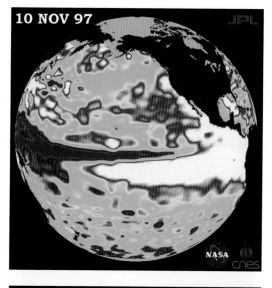

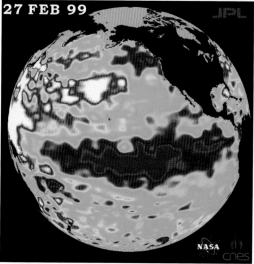

Figure 18a. El Niño conditions in the Pacific Ocean, November 10, 1997. Red and white colors indicate warmer water.

Figure 18b. La Niña conditions in the Pacific, February 27, 1999. Blue and purple indicate cooler water.
NOTE: Images and captions of figs. 18a and 18b were combined as fig. 6 in *Introduction to Air in California* (2006, 25).

TABLE 1

Severity of extreme droughts in the Sacramento and
San Joaquin Valleys

	Sacramento Valley runoff (avg 18.26 MAF/year)*		San Joaquin Valley runoff (avg 5.95 MAF/year)*	
	MAF/year	*% of Avg*	*MAF/year*	*% of Avg*
1929–34	9.78	54	3.33	56
1976–77	6.66	36	1.51	25
1987–92	9.98	55	2.73	46
2012–14	10.50	58	2.51	42

*50-year average in million acre-feet (MAF), from 1961 to 2010.
SOURCE: California Department of Water Resources (2014b).

little supply to meet demand. Different regions may perceive a given year's rainfall totals differently, depending on their local storage capacities, alternative supply sources, and regional populations. The most severe recorded California droughts occurred from 1929 to 1934 and from 2012 to 2014 (extending into 2015 as this book went to print) (table 1). The drought of the early 1930s set the standard that has been applied ever since for developing needed reservoir storage capacity in the state's water system, but the climate is changing and past records may no longer serve as relevant models. The single driest calendar year recorded came in 2014. The 2012 to 2014 drought was the most severe in the last 1,200 years (based on analysis of annual tree-ring growth records). The 2015 snowpack was only 5 percent of the average, the lowest ever measured. In the twenty-first century, climate change appears to be contributing to longer and more intense droughts, not just by reducing precipitation, but also by generating record high temperatures. The year 2014 was

Figure 20. A snow survey site in the Sierra Nevada.

was not "normal" when compared to this longer record; it was, in fact, California's third or fourth wettest century of the past 4,000 years. Scott Stine, in his report on this research, noted, "Since statehood, Californians have been living in the best of climatic times. And we've taken advantage of these best of times by building the most colossal urban and agricultural infrastructure in the entire world, all dependent on huge amounts of water, and all based on the assumption that runoff from the Sierra Nevada will continue as it has during the past 150 years" (1994, 548).

Floods

At the other extreme, floods are equally normal products of the California climate. The Central Valley used to flood annually, becoming a great inland sea when the Sacramento and San Joaquin Rivers, carrying snowmelt from the Sierra Nevada, left their banks to reinhabit the floodplain of the valley floor. Fertile sediments deposited on that flat land were attractive to farmers. Once modern human settlements were established, attempts began to straitjacket the rivers with levees and dikes. Riparian forests were cut down so crops could be planted right up to the edges of the rivers. Towns also were planted, and some grew into major cities. Sacramento, the state's capital, is at the confluence of the American and Sacramento Rivers. Called "River City," it has a long history of floods. Concern about levees and upstream dams remains an issue today. For cities located on floodplains around the state, dams and levees built to prevent floods have often only postponed them.

When warm winter rain events, triggered by "atmospheric rivers," sometimes dubbed the "Pineapple Express," fall onto snow deposited earlier in the winter, rivers suddenly swell. About 40 percent of California's annual snowpack arrives suddenly in just a few atmospheric river events. Water planners and engineers established the "100-year flood" concept not to indicate the actual frequency, but to predict the likelihood of serious floods. Unfortunately, such predictions were based only on the records of flooding available since statehood, in 1850.

California experienced major floods in 1850, 1862, 1955, 1964, 1995, and 1997. The January 1997 event was the largest flood disaster in the state's history (defined by damage to human structures, rather than quantity of runoff water flowing over floodplains). That year 120,000 people were forced from homes and 300 square miles of agricultural land were flooded (fig. 21). Ironically, that flood was followed by a record-setting dry period

Figure 21. Trailer park flooded by the San Joaquin River in 1997.

from February through June 1997. Flooding in 1986 had also marked the beginning of a severe multiyear drought. Those kinds of events make managing for flood protection a challenging trick, one that often conflicts with the maximum storage of water for later use. If water is released during the winter to make room for spring floodwater, the result can be less summer storage.

Since the 1990s, extreme precipitation events in California increased by 20 percent. That may be a consequence of global warming, which is expected to produce wetter and warmer winter storms.

Despite our efforts at controlling them, floods have played an important role in natural water cycling in California, fertilizing floodplains and helping shape the landscape. Water in motion, whether liquid or frozen, also has the power to transform landforms. Glaciers carved broad, U-shaped valleys, such as

Figure 22. U-shaped granite basins scoured by glaciers where the Merced River drops over Nevada and Vernal Falls in Yosemite National Park.

Hetch Hetchy and Yosemite (fig. 22). Rivers have incised narrower, V-shaped canyons, such as the American River canyon in the Sierra Nevada foothills. They also carried debris sediments on down to the Central Valley, building its rich alluvial soils.

Figure 23. Mill Creek in Lundy Canyon.

John Muir listened to the voices of water as it did such work, and wrote of "silvery branches interlacing on a thousand mountains, singing their way home to the sea," or "booming in falls, gliding, glancing with cool soothing, murmuring" (1901, 181, 182) (fig. 23). Mountain streams, Muir said, sang "the history of every avalanche or earthquake and of snow, all easily recognized by the human ear...beside a thousand other facts so small and spoken by the stream in so low a voice the human ear cannot hear them" ([1938] 1979, 95).

Wherever rain falls from the sky over California and rivers sing of their interactions with the land, water speaks in eloquent voices across this state's water landscape.

CHAPTER TWO

California Water Landscape

By such a river it is impossible to believe that one will ever be tired or old. Every sense applauds it. Taste it, feel its chill on the teeth: it is purity absolute. And listen again to its sounds: get far enough away so that the noise of falling tons of water does not stun the ears, and hear how much is going on underneath—a whole symphony of smaller sounds, hiss and splash and gurgle, the small talk of side channels, the whisper of blown and scattered spray gathering itself and beginning to flow again, secret and irresistible, among the wet rocks.

—Wallace Stegner, *The Sound of Mountain Water*

PRISTINE WATERSCAPE

Study a map of California's pristine waterscape (map 4). Some of its major features are startlingly unfamiliar to twenty-first century Californians. The massive Tulare Lake, far bigger than Lake Tahoe, occupies the southern end of the Central Valley. Farther south are the smaller Buena Vista and Kern Lakes. Those three lakes, gone from most of today's maps, are connected by expanses of freshwater marsh, sloughs that flood during wet seasons. The Salton Sea is missing; its basin is marked as a region of saline land that experiences "intermittent water."

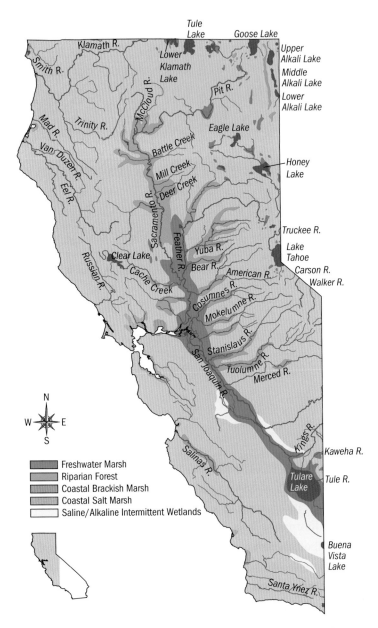

Map 4. California's pristine waterscape (redrawn from Kahrl 1979 and McClurg 2000b).

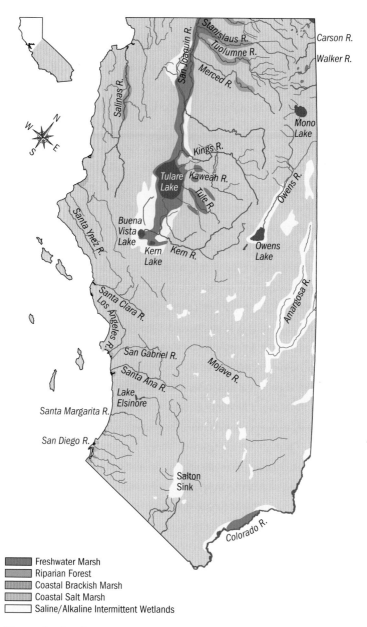

Carson R.

Walker R.

Stanislaus R.

Tuolumne R.

San Joaquin R.

Merced R.

Salinas R.

Mono Lake

Kings R.

Tulare Lake

Kaweah R.

Tule R.

Owens R.

Santa Ynez R.

Buena Vista Lake

Kern Lake

Kern R.

Owens Lake

Santa Clara R.

Los Angeles R.

San Gabriel R.

Santa Ana R.

Amargosa R.

Mojave R.

Lake Elsinore

Santa Margarita R.

San Diego R.

Salton Sink

Colorado R.

N
W E
S

▬ Freshwater Marsh
▬ Riparian Forest
▦ Coastal Brackish Marsh
▨ Coastal Salt Marsh
☐ Saline/Alkaline Intermittent Wetlands

Map 4. *(continued)*

(At least four times between AD 700 and 1580, sediments built up until the Colorado River jumped from its channel and flooded the Salton Sink, creating lakes that gradually evaporated after the river shifted back toward its other terminus in the Gulf of California.) Owens Lake *does* have water—a blue expanse that would be refreshing for contemporary travelers in the Eastern Sierra who pass by today's dry, dusty lakebed. It too would spread over nearby lands during wet cycles. (Mapmakers estimate "natural shorelines," but the surface area of many lakes fluctuated tremendously with California's natural climate variations.) In the northeast corner of the state, Lower Klamath, Tule, and Goose Lakes appear much larger than they would on contemporary maps. All the state's rivers are free flowing, of course, so the map does not show reservoirs later created by dams. A modern map at this scale would show hundreds of reservoirs; one at a larger scale would show thousands.

Two of our major river arteries—the Sacramento and San Joaquin Rivers—gather water from tributaries that drain 37 percent of the state. They merge into a complex of estuary channels that pass water through San Francisco Bay into the Pacific Ocean. Though contemporary maps typically show rivers as solid blue lines following these historic routes, today some stretches have been completely dewatered. For example, since Friant Dam was built in the 1940s, almost no San Joaquin River water has come down as far as the Delta. About 60 miles of its channel are dry, the flow diverted to irrigation (though restoration is underway in 2015). Every river in the arid Tulare Lake basin has dry portions due to diversions into irrigation channels and aqueducts.

Besides the unfamiliar lakes and rivers, the most eye-catching features on the map of the pristine waterscape are more than five million acres of wetlands. Four million of those acres, including both freshwater marsh and riparian forest, sprawl the length of the Central Valley. They do not just parallel rivers, but spread across much of the valley floor. That reflects the winter and spring flooding that once regularly turned the basin (where much of the land is near sea level) into a vast seasonal sea, slowly draining back into the rivers. Today only 350,000 acres of the valley's marshes remain and nine of every 10 acres of its riparian woodlands are also gone. Water that enters most of the remaining wetlands is artificially managed. Natural flooding is a rare reminder that the forces that formed the California waterscape refuse to be fully tamed.

Less obvious on the map, yet totaling hundreds of thousands of acres, marshes also border much of the coastline at the mouths of rivers and around bays and lagoons in both northern and southern California. An extensive freshwater marsh stands out along the west bank of the Colorado River, in striking contrast to the neighboring desert. San Francisco Bay is surrounded by 200,000 acres of brackish and salt marshes (only 35,000 acres still survive today). The surface of the bay itself was also 240 square miles bigger; much open water has succumbed to modern urban landfills (maps 5–7). Another 80,000 acres of marshland border Suisun Bay, where water emerges from the maze of Delta channels and is passed along to San Pablo Bay and San Francisco Bay. Today, what remains of Suisun Marsh is still the largest salt marsh in the lower 48 states, representing 12 percent of the total wetlands acreage left in California.

The Delta, where water spreads into 700 miles of channels, forms the largest estuary on the nation's west coast. It once supported 345,000 acres of tidal marsh but is now down to 8,000 acres. Nearly all Delta marshlands have been transformed into

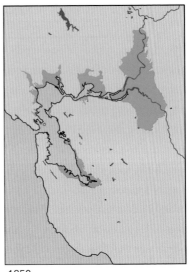

Map 5. San Francisco Bay–Delta
wetlands and urbanization, 1850
(redrawn from U.S. Geological
Survey, n.d.).

1850

Urban Extent
Tidal Wetlands

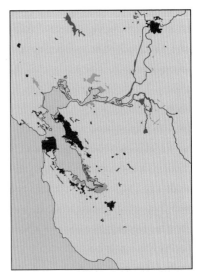

Map 6. San Francisco Bay–
Delta wetlands and urbaniza-
tion, 1940 (redrawn from U.S.
Geological Survey, n.d.).

1940

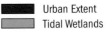

Urban Extent
Tidal Wetlands

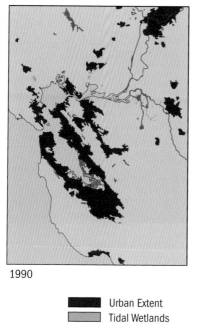

1990

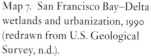
Urban Extent
Tidal Wetlands

Map 7. San Francisco Bay–Delta
wetlands and urbanization, 1990
(redrawn from U.S. Geological
Survey, n.d.).

islands surrounded by levees (fig. 24). Most of the farmland
behind those levees sits below sea level. When peat soils typical
of the Delta were "reclaimed," the dry organic peat oxidized,
carried off as carbon dioxide by the winds that funnel through
the gap in the Coast Ranges. The islands keep sinking, as much
as two inches a year protected only by the levees.

Freshwater that flows into the sea acts as a hydraulic dam that
holds back salt water. The distance that salt water moves inland
with each rising tide is thus determined by the amount of runoff
coming through the Delta towards the bay. The mixing of the
waters determines habitat quality for resident fish, such as Delta
Smelt *(Hypomesus transpacificus)* and Sacramento Splittail *(Pogonich-*
thys macrolepidotus), and for anadromous fish, such as salmon,

Figure 24. Sacramento–San Joaquin Delta channels and islands.

Steelhead Trout *(Oncorhynchus mykiss)*, and Pacific Lamprey *(Entosphenus tridentatus)*, which travel between inland spawning sites and the sea. Historically, during particularly high flow periods, the bay could become mostly fresh. During droughts, salt water could move inland through the Delta all the way to river mouths.

Only about half of the river water that historically passed from the Delta out to sea normally makes it that far today. About 7 MAF of the freshwater reaching the estuary are pumped from the Delta into aqueducts and shipped south. The estuary has become the diversion point for water serving two-thirds of California's population and irrigating San Joaquin Valley and Tulare Basin farms. As a result, salinity is higher in the bay, and the zone where salt and freshwater mix has shifted inland.

Southern California wetlands have also declined or disappeared. Sportsman and author Charles Holder described Southern California coastal marshes, as he knew them in 1906, in his book *Life in the Open* (51, 52):

Figure 25. Upper Newport Harbor, one of the remnant coastal marshes of Southern California.

All the country to the south of the Palos Verde, near San Pedro, and extending to Long Beach, is a shallow back bay, a series of lagunas or canals, often running back into the country to form some little pond or lake. At Alamitos, where the San Gabriel River reaches the sea, and at Balsa Chica...and other places along shore to San Diego we shall find these lagunas, or sea swamps, the home of the duck, goose, and swan. The season begins in November...the air is clear, and the distant mountains stand out with marvelous distinctness...No more beautiful sight than this can be seen in Southern California when these vast flocks pass up and down, silhouetted against the chaparral of the mountain slopes.

Ninety percent of the coastal marsh acreage from Morro Bay to San Diego is gone. Southern Californians can find remnants of it at increasingly rare places, such as Seal Beach and Upper Newport Bay in Orange County, San Diego Bay, and the Tijuana Estuary (fig. 25).

The marshlands that persist are vitally important for many plants, fish, mammals, and breeding and migratory birds. The nutrient mix in coastal estuaries makes them 10 times as productive as the open ocean. Salt marshes serve as nurseries for the majority of oceanic fish and shellfish that are commercially harvested. Fifty percent of the animals listed as endangered or threatened in the United States utilize wetland habitats. Wetlands serve as natural places for floodwaters to spread. Their vegetation controls erosion and cleans water by trapping and filtering sediments.

Wildlife once found in California's marshlands included a half million Tule Elk *(Cervus elaphus nannodes)*, enough Beaver *(Castor canadensis)* to attract trappers long before statehood, and even Grizzly Bears *(Ursus arctos)*, so ubiquitous in the lowlands of California that they became the symbol of the state. Enormous numbers of migrating waterfowl took advantage of the state's wetlands each spring and fall. Sixty million ducks and geese still used Central Valley wetlands in 1950, but by 1996, only three million of the birds traveling the Pacific Flyway found refuge in the remnant valley marshes (fig. 26).

A unique type of wetlands habitat, the vernal pool, was once found on millions of acres in the Central Valley, in Sierra Nevada foothills, in Coast Range valleys, and along the South Coast basins. A vernal pool is a tiny watershed, an ephemeral ecological island, where water collects during the wet season over impermeable hardpan or rock. During the summer, such pools dry up, but tiny cysts and seeds lie dormant in the bottom mud, waiting for another wet cycle. Winter rains awaken fairy shrimp, Tadpole Shrimp *(Lepidurus packardi)*, insects, and plants that are specially adapted to and completely dependent on these environments. Through the spring, as the ponds gradually shrink,

Figure 26. Waterfowl at the Modoc National Wildlife Refuge.

rainbow bands of flowers ring them, segregated by the particular water requirements of each species (fig. 27). Those that need to keep their roots constantly wet grow and bloom closest to the diminishing water body. Farther out are those that prefer drier conditions. Before a pool completely dries, life cycles are rapidly completed. Encysted eggs and seeds settle once again to the bottom mud.

Vernal pools range in size from the 180-acre Table Mountain Lake in Tehama County down to small "hog wallows." About 90 percent of the Central Valley's vernal pools have been lost to plowing and leveling by farmers or to concrete and asphalt. California's list of endangered or threatened species includes five species of fairy shrimp and a tadpole shrimp indigenous to vernal pools. Much of the remaining vernal-pool habitat coexists with cattle ranching along the rim of the Great Valley. The protection of vernal pools became an issue when the University of

Figure 27. A vernal pool, with flowers sorted by their particular needs for water.

California established its newest campus near Merced on land containing 20,000 acres of the now-rare habitat.

The 20,000 miles of rivers and streams in California form 60 major watersheds. Today only one of the state's major river systems, the Smith River on the North Coast, is completely free of dams. The Cosumnes River, in the central Sierra Nevada, is the only major drainage on the west side of the range that still has unregulated flow. Even the Cosumnes has minor dams that pool some of its water to serve diversion intakes (without interrupting all the flow or blocking fish migration).

Eighty California river segments that retain their primitive character have been added to the federal Wild and Scenic Rivers system (map 8). Those segments, totaling 1,900 miles, are protected from dam construction and have management plans to guide activities along their corridors. "Wild" rivers "represent

Map 8. Wild and scenic rivers of California (redrawn from California Department of Water Resources 1998).

vestiges of primitive America"; this formal designation is akin to "wilderness" status. "Scenic" rivers have shorelines that are mostly primitive and undeveloped. "Recreational" status is given to areas easily accessible by roads and with some development. California's own state Wild and Scenic Rivers system prohibits damming and diversion of water and protects the first line of vegetation along rivers. All of California's designated rivers have been incorporated into the federal system.

GROUNDWATER

We naturally focus most on what we can see, but groundwater is not truly separate from surface water. It is another part of the natural water cycle, even if water travels extremely slowly through some groundwater basins. In an aquifer, or water-bearing layer, pores between particles of soil or rock are saturated with water. The top of the saturated region is termed the "water table." Marshes or springs may occur where the water table either intersects the surface or finds a route through rock and is forced up by pressure from below. California's valley floors once had many artesian wells where such free-flowing springs reached the surface.

Groundwater is an important part of California's water supply. If you live in Norco or Visalia, Woodland or Mono City, you are among the Californians—more than one in four—who rely entirely upon groundwater. Half of California residents receive some groundwater through their faucets. About 760,000 acres of farmland are irrigated, in part, with groundwater (fig. 28). Forty percent of the state's average annual water supply comes from wells. That proportion can jump to 60 percent or more during severe drought years.

Figure 28. Irrigation with groundwater.

California has 450 known groundwater basins (map 9). Statewide, they hold 850 MAF of water, 20 times more than the surface water supply, and enough to cover California eight feet
deep. Less than half of that water is usable, however, because of
poor quality and the high costs of pumping it from the ground.
Surface water is found primarily in the northern half of the
state, but groundwater is more evenly distributed across California. It is a valuable local source, because its water is usually
tapped close to where it is used, eliminating the need for long-
distance transport facilities.

The Central Valley holds the largest groundwater basins.
Much of the fresh supply there is found within alluvial deposits
of sand, gravel, and silt washed down from the mountains rimming the valley. Below those freshwater-bearing deposits lies a
thick layer of marine sediments, deposited when the Pacific
Ocean covered that area. Deep drilling sometimes penetrates
into salty groundwater trapped in the marine layer. Overdrafting of the freshwater aquifer has, in places, led to intrusion of
saltwater from below as the overlying pressure dropped.

Because it hides beneath our feet, out of sight, there are a
number of common misconceptions about groundwater:

- California has no "subterranean lakes" or "underground
  rivers." Vast aquifer basins have been identified, but
  within those basins, water simply saturates pores surrounding soil particles.

- Groundwater is not locked away, separate from the rest
  of the water cycle. It is a renewable resource and a
  portion of the interconnected wheel. Though surface
  water and groundwater move at different speeds, with
  most surface water recycling far more quickly, surface

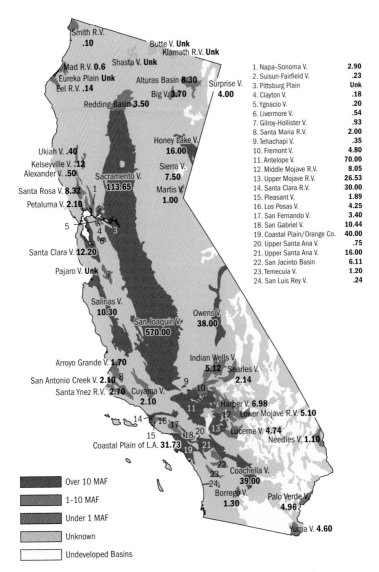

Smith R.V. **.10**
Butte V. **Unk**
Klamath R.V. **Unk**
Mad R.V. **0.6** Shasta V. **Unk**
Eureka Plain **Unk** Alturas Basin **8.30** Surprise V.
Eel R.V. **.14** Big V. **3.70** **4.00**
Redding Basin **3.50**

Honey Lake V.
Ukiah V. **.40** **16.00**
Kelseyville V. **.12** Sierra V.
Alexander V. **.50** Sacramento V. **7.50**
Santa Rosa V. **8.32** **113.65** Martis V.
Petaluma V. **2.10** **1.00**

Santa Clara V. **12.20**

Pajaro V. **Unk**

Salinas V.
10.30 Owens V.
San Joaquin V. **38.00**
570.00

Arroyo Grande V. **1.70** Indian Wells V.
San Antonio Creek V. **2.10** **5.12** Searles V.
Santa Ynez R.V. **2.70** Cuyama V. **2.14**
2.10
Harper V. **6.98**
Lower Mojave R.V. **5.10**
Coastal Plain of L.A. **31.73** Lucerne V. **4.74**
Needles V. **1.10**

Coachella V.
39.00
Borrego V. Palo Verde V.
1.30 **4.96**
Yuma V. **4.60**

1. Napa-Sonoma V.	**2.90**
2. Suisun-Fairfield V.	**.23**
3. Pittsburg Plain	**Unk**
4. Clayton V.	**.18**
5. Ygnacio V.	**.20**
6. Livermore V.	**.54**
7. Gilroy-Hollister V.	**.93**
8. Santa Maria R.V.	**2.00**
9. Tehachapi V.	**.35**
10. Fremont V.	**4.80**
11. Antelope V.	**70.00**
12. Middle Mojave R.V.	**8.05**
13. Upper Mojave R.V.	**26.53**
14. Santa Clara R.V.	**30.00**
15. Pleasant V.	**1.89**
16. Los Posas V.	**4.25**
17. San Fernando V.	**3.40**
18. San Gabriel V.	**10.44**
19. Coastal Plain/Orange Co.	**40.00**
20. Upper Santa Ana V.	**.75**
21. Upper Santa Ana V.	**16.00**
22. San Jacinto Basin	**6.11**
23. Temecula V.	**1.20**
24. San Luis Rey V.	**.24**

Over 10 MAF
1–10 MAF
Under 1 MAF
Unknown
Undeveloped Basins

Map 9. California groundwater basins: known total storage capacity in millions of acre-feet (MAF) (redrawn from Hundley Jr. 2001).

and groundwater do interact. The water that seeps from an aquifer into rivers, lakes, or springs or is pumped out from wells is recharged when precipitation soaks into the ground. Around 7 MAF of water naturally percolate into the state's aquifers each year. Irrigation water seeps into aquifers, too, in the amount of about 6.5 MAF each year. Some water agencies artificially recharge basins by flooding surface ponds or by injecting water deep into aquifers.

· Groundwater is not unlimited. Aquifers can be overdrafted when groundwater is removed faster than it can be replenished. Annual overdrafting averages about 2.2 MAF across the state and about 800,000 AF in the Central Valley alone, but accelerates during droughts.

· Percolation of groundwater through an aquifer cannot clean out any and every contaminant. Though natural filtration makes groundwater generally purer than surface water, our society has learned bitter lessons about its limitations. Specific wells or entire groundwater basins have been lost, essentially forever, to contamination.

HYDROLOGIC REGIONS

California's major drainage basins, which share similar precipitation and runoff patterns, may be grouped into 10 hydrological regions. These water-planning areas are bounded by the crests of the state's mountain ranges. Beginning in the north and working southward (as so much of the water does, these days), they are the North Coast, Sacramento River, North Lahontan, San

Francisco Bay, San Joaquin River, Central Coast, Tulare Lake, South Lahontan, South Coast, and Colorado River regions (map 10). Maps of the state's geomorphic provinces and natural habitat regions look similar to those of its hydrologic regions, because topography influences climate and biological responses. The regions take in a variety of ecosystem types, however, because they include elevations that may range from above the snow line to below sea level. Each region also incorporates the watersheds of many rivers that share similar climates and generally terminate in the same place, either on the coast, at the Bay-Delta, or in inland basins.

North Coast Region

From the Oregon border, the North Coast region extends southward through the Mendocino coast and down to Tomales Bay, north of San Francisco. It reaches inland to the crests of the Klamath Mountains and the Coast Ranges. Major cities in the region include Crescent City and Santa Rosa. Here, the highest rainfall totals in the state create California's version of rain forests, where coast redwoods *(Sequoia sempervirens)* and Douglas-firs *(Pseudotsuga menziesii)* carpet rugged mountain slopes. Much of this watershed receives over 100 inches of rain per year, and some areas near the Oregon border receive almost 200 inches.

Some winter precipitation falls as snow on the Klamath Mountains, but summer's coastal fogs are even more important here than the moderate snowpack of winter. Coast redwoods specialize in growing massive on long, slow drinks of harvested summer fog (fig. 29). You may feel the need for a raincoat when walking through groves of the giant trees while mists drift through their crowns. Fog collects on needles, and the forest

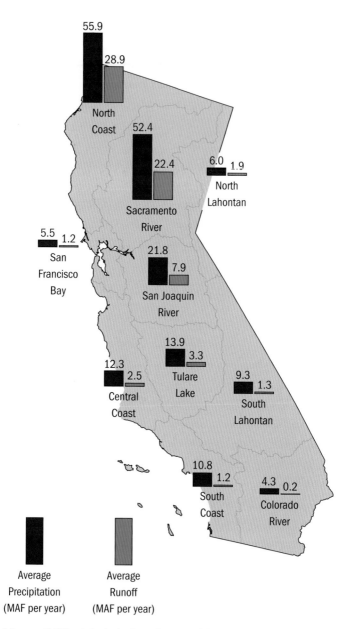

Map 10. California hydrologic regions: precipitation and runoff (redrawn from California Department of Water Resources 1998).

Figure 29. Coast redwoods slowly drinking in the summer fog.

floor is continually splattered by drips that can account for more than 30 percent of the water reaching the ground.

North Coast rivers provide local water supplies; critical spawning habitat for Coho and Chinook Salmon *(Oncorhynchus kisutch* and *O. tshawytscha)*, Steelhead Trout, and other anadromous fish; commercial and sportfishing; and whitewater opportunities. They carry 40 percent of the state's runoff, averaging 28.5 MAF each year. The highest flood peak discharges ever recorded in the state occurred here. North Coast rivers regularly turn brown from sediment loads during intense rain episodes, a natural condition that is exacerbated by logging and grazing practices in the watersheds.

In the far northwest corner of California, close to the Oregon border, the Smith River watershed sends the ocean the highest volumes of runoff water per acre in the state. It also has the proud, but lonely, distinction of being the only watershed of a major river in the state that is entirely free of dams.

The Klamath is California's second-largest river. With headwaters in the mountains of eastern Oregon, the Klamath irrigates farms in southern Oregon before crossing into California. It annually sends over 11 MAF of water toward the Pacific Ocean, draining a 12,100-square-mile watershed that includes Redwood National Park. Portions are protected in the Wild and Scenic Rivers system. Awareness that the Klamath watershed was part of the state's waterscape increased among Californians when requirements for farmers to share water with threatened populations of anadromous fish—Steelhead, Chinook Salmon, Coho Salmon, Cutthroat Trout *(O. clarki)*, Green Sturgeon *(Acipenser medirostris)*, and others—turned into newsworthy standoffs between Oregon farmers and government agency personnel early in the twenty-first century (see the "Raze Existing Dams?" section, Chapter 5).

The 170-mile-long Trinity River drains into the Klamath River, but as much as 90 percent of its water has been diverted entirely out of the North Coast drainage and into the Sacramento Valley. Salmon runs declined as a result of the reduced flows down the Trinity, affecting not only the commercial harvest, but also the subsistence and ceremonial roles the fish had for Hoopa and Yurok Indians. In-stream flow requirements have restored some of the flow down the Trinity's natural channel. Major portions of the Trinity and of the Klamath's other tributaries, the Salmon and Scott, are part of the Wild and Scenic Rivers system.

Humboldt Bay contains the largest wetlands estuary north of San Francisco. From Arcata, on the bay, north of Eureka, the Mad River extends back from the sea 90 miles into the Coast Ranges. South of Eureka, the Van Duzen River passes through Grizzly Creek Redwoods State Park.

Figure 30. The Eel River.

Travelers on Highway 101 cross and recross the winding Eel River as they travel northward through redwood forests from Weott to Fortuna. Most of the Eel is within the Wild and Scenic Rivers system (fig. 30). The peak flood discharge measured on the Eel was 753,000 cubic feet per second. This record exceeds anything that the Sacramento River has generated from its much larger watershed—an indication of the extreme rainfall events in the North Coast region.

The Mattole drains the Lost Coast between Fort Bragg and Cape Mendocino. Runoff from the mountains west of Willits comes down the Noyo through a relatively small coastal watershed, reaching Noyo Harbor at Fort Bragg. The Navarro emerges on the Mendocino coast between the towns of Mendocino and Elk, after just a 19-mile descent from its headwaters.

South of Point Arena, the Gualala River enters the sea. There was once a proposal to suck freshwater offshore from the mouths of the Gualala and the Albion, north of the Navarro, and tow it in massive bags down the coast for sale to San Diego. The idea was dropped in 2003 because of the opposition of local residents and the expense of the environmental documentation required by the California Coastal Commission and the State Water Resources Control Board (SWRCB).

Not far from Fort Ross (a historic Russian settlement established in 1812), the Russian River enters the ocean at Jenner. Its headwaters are in Mendocino County, in the hills above Ukiah. Since 1908, the Russian River has also carried some Eel River water, diverted through a tunnel from a reservoir on the headwaters of the Eel.

Sacramento River Region

The Sacramento River, California's largest, has a 26,548-square-mile watershed that drains the northern half of the Central Valley, with tributaries reaching toward the Oregon border and across the northeast corner of the state. The river carries 31 percent of the state's total runoff, about 22.4 MAF per year. Its watershed includes the eastern slopes of the northern Coast Ranges, Mount Shasta, and the southern Cascade peaks, but it receives most of its runoff from the west slope of the Sierra Nevada. Sacramento River tributaries in the Sierra foothills became the first mining sites of the Mother Lode, where the gold rush began.

Edwin Bryant described the Sacramento River as he saw it in 1846 near Sutter's Fort, where it was "nearly half a mile in width. It is fringed with timber, chiefly oak and sycamore. Grape-vines and a variety of shrubbery ornament its banks, and give a most charming effect when sailing upon its placid and limpid current. I never saw a more beautiful stream. In the rainy season, and in the spring, when the snows on the mountains are melting, it overflows its banks in many places. It abounds in fish, the most valuable of which is the salmon…the largest and the fattest I have ever seen. I have seen salmon taken from the Sacramento five feet in length. All of its tributaries are equally rich in the finny tribe" ([1848] 1985, 271–72).

The Sacramento Valley still has 175,000 acres of wetlands, most managed by private duck clubs in the Butte, Colusa, and American basins and the Delta, but this entire watershed has been transformed by 147 dams. Dams on the river and its tributaries blocked access to most of the historic spawning grounds for four runs of native Chinook Salmon that once migrated by the millions to the upper watershed to spawn. Most of the main river channel has been confined between levees. Flood control and irrigation facilities allowed the Sacramento Valley to be transformed into one of the world's major agricultural regions. More than 2.1 million acres of farmland grow rice, wheat, orchard fruits, alfalfa, and vegetable crops (fig. 31). Farmers and city residents, including the state's legislators, live on a floodplain that used to be inundated each spring.

Sacramento River tributaries include several major rivers and many small creeks. The McCloud and Pit Rivers enter at Shasta Reservoir. The Pit River's 5,000-square-mile watershed spreads across much of the northeastern part of the state. Goose Lake, large but shallow, straddles the Oregon–California border; it drained toward the Pit River whenever it was full.

From Mount Lassen, Mill Creek flows down to the Sacramento River. Mill is one of the last streams in the northern Sierra that provide pristine spawning habitat for salmon and Steelhead. A few of the Sacramento's other small tributaries—Battle Creek,

Figure 31. Sacramento Valley rice fields.

Antelope Creek, Deer Creek, Big Chico Creek, and Butte Creek—also remain undammed.

The Feather River is the primary water source for the State Water Project (SWP). The middle fork of the Feather, upstream from Oroville Dam, is part of the Wild and Scenic Rivers system. The Yuba and Bear River tributaries join the Feather before it reaches the Sacramento.

On the south fork of the American River, in 1848, James Marshall made a discovery that would shape much of California's early history and many environmental changes. In May of that year, entrepreneur Sam Brannan spread the news in the streets of San Francisco, shouting, "Gold! Gold! Gold from the American River!" The river had been named the Rio de los Americanos in 1837, because American trappers were frequenting that part of Mexican Alta California. It now passes through Sacra-

Figure 32. Salmon fishermen and rafters on the American River above Sacramento.

mento along a 26-mile river parkway (fig. 32). The upper north fork of the American River is part of the Wild and Scenic Rivers system.

One of California's largest natural lakes, Clear Lake, is in the Coast Ranges within the Sacramento River region. With a 100-mile shoreline and 43,000 surface acres, Clear Lake is the largest freshwater lake inside California's borders (Lake Tahoe is quite a bit larger but is partly in Nevada). It is known as the Bass Capital of the West for its recreational fishing, but its age is a more venerable distinction: the lake has been there for about three million years, making it one of the oldest in North America. Water from Clear Lake reaches Cache Creek and drops into Yolo County in the Central Valley.

North Lahontan Region

The North and South Lahontan regions are named for Lake Lahontan, the Ice Age lake that once covered 8,665 square miles of the Great Basin, with tributary streams in the Eastern Sierra. Much of the runoff from the Eastern Sierra Nevada drains into terminal lake basins in Nevada that are the isolated remnants of Lake Lahontan. The North Lahontan region extends from the Oregon border south to the Walker River watershed above Bridgeport Valley.

Lake Tahoe is Northern California's largest water feature and the dominant feature of a hydrologic region that is profoundly influenced by the Sierra rain shadow. Tahoe, between four and five million years old, is the oldest lake in North America. It is 22 miles long and 12 miles wide. The state line splits it from north to south, with a jog to the east in the middle of the lake that turns the boundary line to parallel the Sierra crest. The lake is 1,685 feet deep, so it never freezes, even though its surface is 6,229 feet above sea level.

An early tourist to Lake Tahoe, Mark Twain, marveled at its clarity in 1861 (fig. 33): "So singularly clear was the water, that where it was only twenty or thirty feet deep the bottom was so perfectly distinct that the boat seemed floating in the air! Yes, where it was even *eighty* feet deep. Every little pebble was distinct, every speckled trout, every hand's-breadth of sand. So empty and airy did all spaces seem below us, and so strong was the sense of floating high aloft in nothingness, that we called these boat-excursions 'balloon-voyages'" ([1872] 1972, 168–69).

As beautiful as it remains, Tahoe's clarity has been decreasing. In the 1960s, white disks used by researchers to monitor water visibility could be seen 100 feet below the surface. The disks disap-

Figure 33. Lake Tahoe.

pear at 70 feet now, indicating that in recent decades visibility has diminished by about one foot per year. Runoff from homes, casinos, golf courses, and the streets and septic systems that service them washes nutrients into Lake Tahoe (fig. 34). Nitrogen and phosphorus fertilize algae growth in the water, which had been renowned for its near sterility. This eutrophication has been exacerbated by the loss of adjacent marshes that used to filter runoff.

Figure 34. Runoff from surrounding development entering Lake Tahoe, creating eutrophication.

Lake Tahoe's only outflow is the Truckee River. Its watershed in California drains to Winnemucca and Pyramid Lakes in Nevada. Cutthroat Trout once climbed the Truckee to spawn, coming all the way to Lake Tahoe. That population went extinct after irrigation diversions interfered with flows needed by the fish. A Cutthroat fishery was reestablished with fish brought from Walker Lake, farther south in Nevada. Another native fish *(Chasmistes cujus)*, a sucker named Cui-ui by the local Paiute Indians, became an endangered species. In 1999, after decades of legal disputes, an allocation of water between California and Nevada provided water to protect Pyramid Lake and wildlife refuges farther downstream.

The Carson and Walker Rivers drain Eastern Sierra watersheds that also send water to wetlands and lakes in western Nevada. From headwaters near Sonora Pass and above Bridgeport, the

Walker flows northward, then turns south to terminate in Walker Lake. Diversions for agriculture after the river leaves California have caused Walker Lake to drop 140 feet since 1900. Salinity levels in the lake have increased to near the limit for freshwater fish, threatening the Cutthroat Trout and Tui Chub *(Gila bicolor)* populations. Endangered White Pelicans *(Pelecanus erythrorhynchos)* and Common Loons *(Gavia immer)* will lose an essential migration stop unless California and Nevada water rights holders begin sharing with the lake.

San Francisco Bay Region

Cool and foggy on the coast, but with inland hills that experience the hot, dry summer weather of Mediterranean climates, the San Francisco Bay region takes in watersheds that drain into San Francisco, San Pablo, and Suisun Bays. The region's eastern boundary runs from the confluence of the Sacramento and San Joaquin Rivers along the watershed crests of the Coast Ranges, including the Berkeley and Oakland hills in the East Bay (fig. 35). In the north, the region takes in the Napa River valley and Tomales Bay. Where the San Andreas Fault enters the Pacific Ocean, Lagunitas Creek supplies the Tomales Bay estuary with freshwater gathered from Mount Tamalpais and other Marin County hillsides. Coho Salmon and Steelhead Trout in Lagunitas have declined, as have conditions for the endangered California Freshwater Shrimp *(Syncaris pacifica)*. The region extends down the San Francisco Peninsula to the Santa Cruz Mountains.

Urbanization and industrial development have transformed the Bay Area wetlands and the local watersheds. Enormous growth in San Francisco, Oakland, San Jose, and their sprawling

Figure 35. Egret at the Martinez Regional Shoreline, where the Carquinez Strait connects San Pablo Bay and San Francisco Bay.

suburban satellites was only possible because of water imported from outside the local watersheds.

San Joaquin River Region

The San Joaquin Valley is drier and hotter than the Sacramento Valley, but more than 100,000 acres of wetlands persist there, mostly managed by private duck clubs (fig. 36). The region is bounded on the west by the Diablo Range portion of the Coast Ranges and on the east by the Sierra Nevada crest. There are five million acres of irrigated farmlands in the San Joaquin Valley; Fresno and Tulare Counties are the top agricultural revenue-generating counties in the nation. Crops include cotton, corn, grains, grapes, vegetables, orchard fruits, nuts, citrus, and alfalfa (fig. 37). Water rights developed for agriculture have made

Figure 36. Geese at the San Joaquin River National Wildlife Refuge near Los Banos.

Figure 37. Harvesting peas in the Central Valley.

the land valuable to builders, however. Fertile farmland is rapidly succumbing to urban sprawl, repeating a pattern that completely transformed the nation's former number one agricultural region, Los Angeles County.

Groundwater is a major supply source in the San Joaquin Valley, for both agricultural and urban use. Portions of the region have experienced severe land subsidence due to groundwater overdrafting. Heavy pumping from lands between the Mokelumne and Stanislaus Rivers east of Stockton also caused poor-quality Delta water to migrate toward city wells.

The San Joaquin River watershed drains nine percent of the state's runoff water, about 6.4 MAF in an average year. Its upper portion includes the west slopes of Mammoth Mountain, Devil's Postpile National Monument, and parts of the Ansel Adams Wilderness. Once the river reaches the valley, it turns sharply northward and picks up water from major tributaries as they emerge from the mountain foothills. Because about 60 miles of the San Joaquin were dewatered when Friant Dam was built, its water was diverted to farm irrigators, the only San Joaquin River water that has entered the Delta since the 1940s was contributed by its tributaries, supplemented by farm return drainage water plus treated urban sewage. A restoration plan is in place and flows to restore salmon migration were supposed to be reestablished in 2015.

Two of the rivers in this region, the Cosumnes and the Mokelumne, enter the Delta north of Stockton without actually joining the San Joaquin. The Cosumnes, one of the smallest Sierra Nevada rivers, is the only watershed in the western Sierra with no major dam (some small dams serve local irrigation but do not block fish passage). Without flood control structures, the river can behave naturally. In the Cosumnes Nature Preserve on the valley floor, old levees have been purposely breached, allowing the river to seasonally leave its banks and reoccupy parts of its natural floodplain. The Mokelumne River serves as the major water supply for the East Bay Municipal Utility District (EBMUD) and is also diverted near Lodi for irrigation.

The major tributaries to the San Joaquin are, from north to south, the Stanislaus, Tuolumne, and Merced Rivers. The Stanislaus drains the Sonora Pass area. Its north fork passes through Calaveras Big Trees State Park. In the 1970s, national attention became focused on New Melones Dam, whose reservoir flooded high-quality whitewater-rafting segments of the Stanislaus River canyon.

The Tuolumne River is the largest tributary to the San Joaquin. Its headwaters are in Yosemite National Park. Snow that melts west of Tioga Pass, the eastern entrance station to the park, flows through Tuolumne Meadows, drops down dramatic cascades into the Grand Canyon of the Tuolumne, then enters Hetch Hetchy Valley. John Muir called Hetch Hetchy "the 'Tuolumne Yosemite,' for it is a wonderfully exact counterpart of the Merced Yosemite, not only in its sublime rocks and waterfalls but in the gardens, groves, and meadows of its flowery parklike floor" (1912, 187).

Today, the city of San Francisco's O'Shaughnessy Dam floods Hetch Hetchy Valley; farther downstream, New Don Pedro Reservoir diverts water to irrigation districts. There has been a severe decline in Chinook Salmon in the Tuolumne River. Fewer than 100 fall-run salmon returned to the river during 1991 and fewer than 200 in 1992, compared to a historical maximum of 130,000 in 1944. Numbers rose to 17,873 in 2000, but dropped again to 211 in 2007 despite habitat restoration efforts.

Perhaps the most famous 2,400-foot stretch of water flowing to the Merced River is Yosemite Creek's two-stage drop over

Figure 38. Yosemite Falls.

Yosemite Falls (fig. 38). Portions of the Merced may be the most visited river stretches in California, as three million tourists come to Yosemite Valley each year. Many swim and float in the Merced as it meanders through the valley, and even more hike to its waterfalls. On the main channel of the Merced, water cascades over 600-foot Nevada Falls and 317-foot Vernal Falls. Today the lower Merced River has the southernmost run of Chinook Salmon on the West Coast.

Central Coast Region

Central Coast rivers originate in the Coast Ranges and drain to the Pacific Ocean. The region extends from Santa Cruz to Santa Barbara. Major rivers include the Carmel and the Big Sur, with headwaters in the steep mountains of the Ventana Wilderness, inland from the Big Sur coast. The Big Sur is part of the Wild and Scenic Rivers system. It was once a major spawning ground and nursery stream for Steelhead Trout, with up to 3,000 spawners per year, but dams, diversions, and groundwater pumping along the river reduced its flows and the number of fish that successfully spawn.

The Salinas River makes a long northwesterly run to Monterey Bay. Much of the water used for agriculture in the Salinas Valley comes from groundwater replenished by the river. In *East of Eden*, John Steinbeck described the Salinas as "only a part-time river. The summer sun drove it underground. It was not a fine river at all, but it was the only one we had and so we boasted about it" ([1952] 1995, 4) (fig. 39). A tributary of the Salinas, the Arroyo Seco River sustains a small population of threatened Steelhead Trout that migrate there from the ocean to spawn. Downstream of the Los Padres National Forest boundary, the Arroyo Seco floodplain waters one of the largest native sycamore forests in central California.

San Luis Obispo Creek flows out of the Santa Lucia Range, through the town of San Luis Obispo, and out to sea. Though streamflow below the town is primarily treated wastewater in the

Figure 39. The Salinas River.

dry season, the drainage still supports populations of Southwestern Pond Turtles *(Clemmys marmorata pallida)*, Red-legged Frogs *(Rana aurora draytonii)*, and one of the southerly races of Steelhead.

The Santa Ynez River drains the rugged mountains of the same name north of Santa Barbara. The upper Santa Ynez River has Southern California's longest stretch of free-flowing river accessible to recreational users. Historically, the Santa Ynez supported the largest run of Steelhead Trout in Southern California, but the too common story of reduced flows, high temperatures, and barriers to migration also occurred in this drainage.

Tulare Lake Region

The southern third of the Central Valley contains the Tulare Lake basin. The Tehachapi Mountains form the region's southern boundary. Here the valley floor receives less than 10 inches of rain; Bakersfield averages only six inches of winter rain. Tulare Basin rivers never reached the sea. The Kern River once terminated in Buena Vista Lake, and the Tule, Kings, and Kaweah Rivers (from the Sequoia National Park area) once formed Tulare Lake. All of those rivers have stretches below the foothills that are now completely dewatered. James Carson, in 1852, wrote about

the placid blue water of the Tulare Lake, whose ripples wash the foot of the low hills of the Coast range.... [Tulare Lake] is about fifty miles in length by thirty in width... Buena Vista Lake is a beautiful sheet of water, twenty miles long, and from five to ten in width; it lays nestled in the head of the valley... The slough connecting the Tulare and Buena Vista Lakes is about eighty miles in length... Thousands of wild horses subsist on the grasses growing there now... Every beast and bird of the chase and hunt is to be found in abundance on the Tulares. Horses, cattle, elk, antelope, black tail and red deer, grizzly and brown bear, black and grey wolves, coyotes, ocelots, California lions, wildcats, beaver, otter, mink, weasels, ferrets, hare, rabbits, grey and red foxes, grey and ground squirrels, kangaroo rats, badgers, skunks, muskrats, hedgehogs, and many species of small animals...; swan, geese, brant, and over twenty different...ducks...in countless myriads from the first of October until the first of April, besides millions of...crane, plover, snipe, and quail. (65, 66, 68, 69, 76, 80)

Carson was describing the largest single block of wetland habitat in California, approximately 500,000 acres. Currently there are only about 6,400 acres of flooded wetland habitat left in the basin, mostly in the Kern and Pixley National Wildlife Refuges. Tulare and Buena Vista Lakes have vanished from maps of the southern Central Valley, their river tributaries diverted and the lakebeds transformed into farm fields (fig. 40). Still, in about one-third of winters the basin experiences

Figure 40. Cotton harvest in the Tulare Basin.

flooding. Lakebed farmers must also contend with salt buildup in the soil. That is something that all irrigators must address, but it is exacerbated in the Tulare Basin, because the topsoil sits above an impermeable layer that does not allow drainage. Salts increase unless they are flushed away with lots of water. The wastewater settles into sinks, where toxic chemicals can become a hazard for wildlife.

Upper portions of the Kings River are in Kings Canyon National Park and were added to the Wild and Scenic Rivers system in 1987. The Kings carves away at one of the deepest canyons in North America as it drops more than 13,000 feet. In the valley, it is diverted into a system of irrigation channels and only deposits water on the Tulare Lake bed in extremely wet years.

The Tule River, a tributary of the Kings, has headwaters in the Golden Trout Wilderness of the southern Sierra. One fork passes through the recently designated Sequoia National Monument, east of Porterville.

Figure 41. Golden Trout, the California state fish.

Snowmelt from Mount Whitney, in Sequoia National Park, forms part of the water entering the Kern River. The upper portions of the Kern are part of the Wild and Scenic Rivers system. The south fork and Golden Trout Creek have populations of Golden Trout *(Salmo aguabonita)*, the only native trout found in the southern Sierra (fig. 41). California's largest lowland riparian forest is found along the south fork, upstream from Lake Isabella. The American Bird Conservancy has identified Kern River Preserve, also on the south fork, as a Globally Important Bird Area; it attracts over 300 species of birds, including Yellow-billed Cuckoo *(Coccyzus americanus occidentalis)* and Willow Flycatcher *(Empidonax traillii)*. Stretches of the Kern River below Isabella Dam are popular for whitewater rafting (but flows vary according to releases from the reservoir). Downstream diversions cause the Kern to go dry, but efforts are being made to put water back into the river as it passes through Bakersfield, using groundwater from local wells.

South Lahontan Region

At the Sierra crest, snow and ice loom over a sudden plummet into eastern high desert basins. This region contains the highest and lowest points in the lower 48 states: Mount Whitney, at 14,495 feet above sea level, and not far away, Death Valley, 282 feet below sea level. Owens Valley is a 90-mile-long trough nestled between the Sierra Nevada and the White and Inyo mountain ranges. The Owens River carries snow water the length of the valley to Owens Lake. Farther north, in a separate drainage basin, five smaller streams terminate at Mono Lake, where, according to John Muir, "spirit-like, our happy stream vanishes in vapor, and floats free again in the sky" (1961, 71). The Mojave Desert dominates the southern parts of the region. Using water imported from the Sacramento Valley, cities like Lancaster and Palmdale in Los Angeles County, and Victorville and Apple Valley in San Bernardino County, have grown tremendously in the past few decades. The populations of Inyo and Mono Counties, in the Eastern Sierra, are essentially stable as aqueduct facilities export water from that region to Los Angeles.

Mono Lake is a vast inland sea that supports migratory and nesting bird populations attracted by the lake's enormous productivity. Harsh alkaline water, almost three times as salty as the ocean, provides nutrients to algae, brine shrimp, and Alkali Flies *(Ephydra hians)*. The ancient lake, at least one million years old, has been declared an Outstanding Natural Resource Water by the Environmental Protection Agency and a Globally Important Bird Area by the American Bird Conservancy. Its unique chemical "recipe" has produced a unique ecosystem, a good example of the critical role water plays in shaping plant and animal responses. It is a "three-salt lake," containing carbonates, chlo-

Figure 42. Tufa towers at Mono Lake.

rides, and sulfates. To mimic the mixture requires table salt, baking soda, and Epsom salts in the right proportions. Because carbonates predominate, photogenic limestone tufa towers form where freshwater springs (bearing calcium) enter the carbonate-rich lake (fig. 42).

Mono Lake became the center of a major controversy after decades of stream diversions to Los Angeles caused it to shrink. To stop the decline before the lake became lethally concentrated, a court directed the SWRCB to amend Los Angeles's water licenses. Since 1941, four of the lake's tributary streams, Lee Vining, Rush, Walker, and Parker Creeks, had been completely dewatered by diversions into the Los Angeles Aqueduct system. That violation of state fish and game law was also corrected by the 1994 SWRCB plan implementing the court order.

In 1913, three decades before the Los Angeles Aqueduct system tapped into Mono Basin streams, water from the Owens

River had begun flowing to Los Angeles. Owens Lake dried up within a decade, and the agricultural valley was transformed into a "wholly owned subsidiary" of the city of Los Angeles (figs. 43, 44). Dust storms blowing off the desiccated lakebed began violating state air quality standards, forcing the city to put a portion of the Owens River water back into the lakebed as the twenty-first century opened.

Water in the north fork of the Owens River, at 9,500 feet above sea level, was the best John Muir ever found, he wrote in 1901: "It is not only delightfully cool and bright, but brisk, sparkling, exhilarating, and so positively delicious to the taste that a party of friends I led to it twenty five years ago still praise it, and refer to it as 'that wonderful champagne water;' though, comparatively, the finest wine is a coarse and vulgar drink" (185).

Although trout fishing is today a major recreation activity in Eastern Sierra streams and lakes, there were no native trout in any of these waters; all have been introduced. The only fish in the Owens River system were the Owens Sucker *(Catostomus fumeiventris)*, which used the river as far up as Convict Lake; the small Speckled Dace *(Rhinichthys osculus)*; and the Owens Pupfish *(Cyprinodon radiosus)*.

The dace and pupfish are also found in the Amargosa River, which flows southward from Nevada into California, then turns west into the Mojave Desert and, finally, north into Death Valley. It flows intermittently, but a 20-mile segment near the town of Shoshone is perennial. "Amargosa" is Spanish for "bitter water," but the water has critical importance to desert wildlife and plants. Its riparian habitat supports endangered and threatened birds, including the Willow Flycatcher, Yellow-billed Cuckoo, and Least Bell's Vireo *(Vireo bellii pusillus)*. The Amargosa Vole *(Microtus californicus scirpensis)* is found nowhere else in the world.

Figure 43. Owens Lake before the Los Angeles Aqueduct began diversions.

Figure 44. The bed of Owens Lake today.

The Mojave River drops out of the San Bernardino Mountains and flows eastward, going underground as it enters the desert. The river and the Mojave groundwater basin act as one interacting water source.

South Coast Region

Rivers and creeks in the Southern California coastal basins drain into the Pacific Ocean. This is the West Coast's most populated, most urbanized region. While more than 50 percent of California's population lives here, the area naturally receives less than two percent of the state's precipitation. It would be unable to support even a significant fraction of its population without water imported from other regions of California and the Colorado River. The hydrologic region is bounded in the north by the Santa Barbara–Ventura County line and the San Gabriel and San Bernardino Mountains. The eastern boundary is formed by the San Jacinto and Santa Rosa Mountains, then the Peninsular Ranges down to the Mexican border.

Rainfall across southern California is quite variable. Seasonal averages depend on elevation and topography. Los Angeles gets 15 inches of rain downtown but only eight inches along the coast. At the summit of the San Gabriel Mountains, 40 inches may fall. Groundwater basins were once so full that they produced artesian wells, particularly around San Bernardino and in northwestern Orange County (fig. 45). The basins became badly overdrawn early in the twentieth century.

So much of the watershed has been covered by concrete, asphalt, and buildings that stormwater does not percolate into the ground but rapidly runs into channels and drains. That not

Figure 45. Artesian well near San Bernardino, about 1900.

only reduces groundwater recharge but also carries trash and toxic chemicals to the ocean and beaches.

Much of the Southern California population seems unaware of the remnants of the former riparian corridors, and equally unaware that those local channels once supported populations of Steelhead and Pacific Lamprey, suckers, native frogs, and Southwestern Pond Turtles. The rivers of the Southland are typically small, ephemeral, and intermittent, but, ironically, many of them carry *more* water today than they did before imported water arrived in great quantities in Southern California. Wastewater from millions of people must go somewhere after it passes through treatment plants. "Effluent-dominant" rivers that would naturally be dry except during the rainy season now run all year with treated wastewater.

The 51-mile-long Los Angeles River has flowed mostly through 470 miles of concrete channels and drains, once it drops from headwaters in the San Gabriel Mountains, Santa Susanna

Mountains, Simi Hills, and Santa Monica Mountains. A major groundwater basin in the San Fernando Valley also feeds the river. It enters the ocean at the Los Angeles–Long Beach harbor, through a three-mile-long estuary. Winter rains, however, produced regular flooding, when the river would romp across the floodplain. Now and then, it completely changed its outlet, draining to the ocean through Ballona Creek.

William Mulholland, the city water engineer who oversaw the creation of the Los Angeles Aqueduct, began his career tending water ditches along the river. "The Los Angeles River was the greatest attraction," he told an oral historian in 1931. "It was a beautiful, limpid little stream with willows on its banks" (Spriggs 1931, 67). Since the late 1990s, energy began building aimed at revitalizing the Los Angeles River. The city adopted a master plan in 2007, and in 2011 the river was chosen as one of a few in a federal urban waters pilot program. The Army Corps of Engineers completed an ecosystem restoration feasibility study in 2014. Mulholland would likely be pleased.

To the south, the San Gabriel River enters the ocean at Seal Beach and Los Alamitos Bay, north of the Santa Ana River. The Santa Ana's headwaters in the San Bernardino Mountains are still free flowing, but it is channelized for most of its length through Orange County (fig. 46).

From southwestern Riverside County, the Santa Margarita River heads south for San Diego County, then bends westward and passes through the U.S. Marine Corps base at Camp Pendleton to the Pacific Ocean. One of the last free-flowing rivers in Southern California, the 27-mile-long Santa Margarita is a remnant treasure in this region that has been so urbanized and transformed.

There are amazing waterfalls in eastern San Diego County on the upper San Diego River and its tributary, Cedar Creek.

Figure 46. The channelized Santa Ana River.

The San Diego Mission sits high on a hill overlooking the historic channel of the San Diego River, but the city now occupies much of the floodplain.

Colorado River Region

In the dry southeast corner of the state, an insignificant amount of drainage from California enters the Colorado River, which gets the overwhelming portion of its water from watersheds far to the north in the Rocky Mountains. The Colorado River is 1,440 miles long. Its watershed covers one-twelfth the area of the lower 48 states, passing through parts of Wyoming, Colorado, Utah, New Mexico, Nevada, and Arizona (fig. 47). It forms the border between California and Arizona and finally heads toward the Gulf of California in northern Mexico. The mighty Colorado only reaches the gulf in very wet years, because of the cumulative diversions north and south of the border.

Figure 47. The Colorado River at Picacho State Park.

Naturally high sediment loads give the river a reddish color, which is responsible for its Spanish name. Before Colorado River water is used for domestic purposes, it is often blended with purer sources to reduce salt concentrations.

The region's other dominant water body is the Salton Sea, which looms large on California maps. Even bigger than Lake Tahoe, the Salton Sea stretches across 35 miles of desert and covers 360 square miles. It is 228 feet below sea level. An accidental re-creation of natural flood events that occurred many times, the sea has become a key alternative for the lost waterfowl habitat along the Southern California coast and in the Colorado River delta. Millions of birds of over 400 species visit it each year. Farm runoff keeps it full but also aggravates salinity issues. The sea suffers from a constant salt influx from the Colorado River aggravated by added salts and fertilizers from farm runoff. It is approaching salt levels at which its prolific fish populations may no longer survive. Salinity issues will increase as plans to fallow farmland and divert water to Southern California cities are carried out. Though the farm drainage water increases the salinity of the sea, diverting water to the coast will accelerate the concentration process, reducing water needed to replace evaporation (see the "Asking Too Much of the Colorado River and the Salton Sea" section, Chapter 4).

Today only about 36 percent of the state's surface water still keeps river and riparian ecosystems functioning. Most water moves across California in man-made channels that have facilitated population growth and development. The plumbing has become so thorough that today, almost all Californians are closely connected to one another through their water pipes.

CHAPTER THREE

The Distribution System

The water I will draw tomorrow from my tap in Malibu is today crossing the Mojave Desert from the Colorado River, and I like to think about exactly where that water is. The water I will drink tonight in a restaurant in Hollywood is by now well down the Los Angeles Aqueduct from the Owens River, and I also think about exactly where that water is: I particularly like to imagine it as it cascades down the 45-degree stone steps that aerate Owens water after its airless passage through the mountain pipes and siphons.

—Joan Didion, *The White Album*

EXPANDING WATERSHEDS

Stand at the top of a snow-covered ski run on the summit of Mammoth Mountain in the winter, and think about where the water under your feet may be by midsummer. Much of what drains eastward will end up in Los Angeles instead of Owens Lake. Turn around for a view of the western slope; there water will run down to the San Joaquin River. Some of the river water may reach the Central Valley and turn northward toward San Francisco Bay. Most of it, however, will be diverted onto farm fields in the San Joaquin Valley.

Figure 48. Spring snow conditions near Tioga Pass, west of the Yosemite National Park entrance.

Or get out of your car at Tioga Pass, at the eastern entrance to Yosemite National Park. Melting snow that drains westward from here will feed the Tuolumne River, which merges with San Joaquin River water flowing toward the sea (fig. 48). But much of the Tuolumne water will go, instead, into San Francisco's water supply system. Walk just a few steps outside the park and the drainage begins sloping to the east. Snowmelt from here may reach Lee Vining Creek, a tributary to Mono Lake, but some will be diverted and become the northernmost water entering the Los Angeles Aqueduct.

Drive Interstate 80 from Reno, Nevada, toward San Francisco. The American and Yuba Rivers flow westward, paralleling this route through the mountains, also aiming for the Pacific Ocean via the Golden Gate. Much of that water will be diverted out of the Delta and moved southward to irrigate crops in the San Joaquin Valley.

Farther north, the Feather River is the source for the State Water Project's (SWP) California Aqueduct. Much of that water will emerge from Southern California faucets after traveling over 600 miles through natural channels and aqueducts. Some water brought to Southern California cities covers even greater distances, starting 1,400 miles away in the Colorado River headwaters in Colorado and Wyoming. Winter snowfall in the Sierra Nevada is more critical than local rainfall to the Bay Area's water supply, and both Sierra Nevada and Rocky Mountain winters are far more critical than local weather to Southern Californians.

Because of long-distance aqueducts, the great majority of Californians live within "virtual watersheds," with distant snowpacks and long-distance transportation systems providing most of their water (map 11). Sierra Nevada and Rocky Mountain snow assures the economic wealth of California, a state with the world's fifth largest economy, bigger than almost every *nation's*. The enormous population growth of Southern California and the San Francisco Bay Area was only made possible by damming distant rivers and importing their water, so that local resource limits could be made irrelevant. Without outside water, there could only be about three million people in Southern California, where 18 million now reside from Ventura down to San Diego.

California's water landscape has been reengineered so that roughly 75 percent of the *demand* for water originates south of Sacramento, although 75 percent of the water *supply* in the state comes from north of the capital city. To create twenty-first century California, an intricate network has been engineered across the face of the state (map 12). Of the 42 MAF of "developed water"—water that has been gathered behind dams, pumped from the ground, and transmitted along aqueducts and through pipes—about 80 percent irrigates the state's farms, and the rest

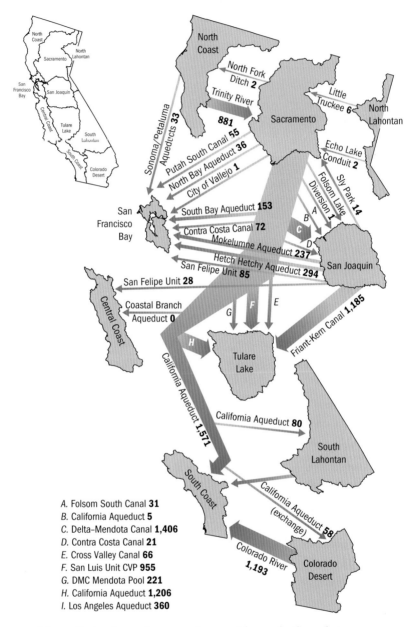

North Coast

North Coast

North Fork Ditch 2

Trinity River

881

Sacramento

Little Truckee 6

North Lahontan

Sonoma/Petaluma Aqueducts 33

Putah South Canal 55

North Bay Aqueduct 36

City of Vallejo 1

Echo Lake Conduit 2

Sly Park 14

Folsom Lake Diversion 1

San Francisco Bay

South Bay Aqueduct 153

Contra Costa Canal 72

Mokelumne Aqueduct 237

Hetch Hetchy Aqueduct 294

San Felipe Unit 85

A
B
C
D

San Joaquin

San Felipe Unit 28

Central Coast

Coastal Branch Aqueduct 0

G F E

H

Tulare Lake

Friant-Kern Canal 1,185

California Aqueduct 1,571

California Aqueduct 80

South Lahontan

South Coast

California Aqueduct (exchange) 58

Colorado River 1,193

Colorado Desert

A. Folsom South Canal **31**
B. California Aqueduct **5**
C. Delta–Mendota Canal **1,406**
D. Contra Costa Canal **21**
E. Cross Valley Canal **66**
F. San Luis Unit CVP **955**
G. DMC Mendota Pool **221**
H. California Aqueduct **1,206**
I. Los Angeles Aqueduct **360**

Map 11. Regional water imports and exports (thousands of acre-feet per year) (redrawn from California Department of Water Resources 1998).

1. Lake Sonoma
2. Lake Mendocino
3. Indian V. Res.
4. Lake Berryessa
5. New Melones Lake
6. New Don Pedro Lake
7. Lake McClure
8. California A.
9. San Vicente Res.
10. Henshaw Res.

State
Federal
Local

Tule Lake
Clear Lake
Trinity Lake
Shasta Lake
Whiskeytown Lake
Red Bluff Diversion Dam
Corning Canal
Tehama-Colusa Canal
Black Butte Res.
Lake Almanor
Stony Gorge Res.
Glenn-Colusa Canal
East Park Res.
Lake Oroville
Clear Lake
New Bullards Bar Res.
Englebright Res.
Folsom Lake
Lake Tahoe
North Bay A.
Folsom South Canal
Camanche Res.
Contra Costa Canal
Mokelumne A.
South Bay A.
Hetch Hetchy A.
Santa Clara Conduit
Grant Lake
Delta-Mendota Canal
Lake Crowley
Hollister Conduit
Madera Canal
San Luis Res.
Millerton Lake
San Luis Canal
San Antonio Res.
Pine Flat Lake
Coalinga Canal
Friant-Kern Canal
Nacimiento Res.
Lake Kaweah
Success Lake
Isabella Lake
Twitchell Res.
Los Angeles A.
Cross Valley Canal
Cachuma Res.
Lake Casitas
Castaic Lake
Silverwood Lake
Lake Mathews
Lake Perris
Colorado River A.
San Diego A.
Coachella Canal
Lower Otay Res.
All American Canal

Map 12. Major water transport systems (redrawn from California Department of Water Resources 1998 and from McClurg 2000b).

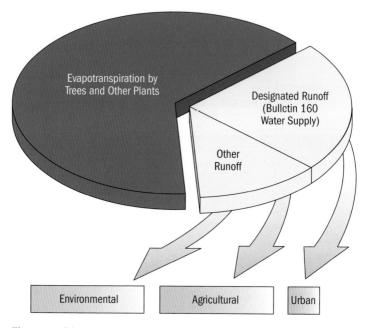

Figure 49. Disposition of California's average annual precipitation (redrawn from California Department of Water Resources 1998).

"irrigates" the urban population and its industrial needs (fig. 49). California's water systems, federal, state, and local, have allocated a total maximum allowable use of 370 MAF of surface water—*more than five times the 70 MAF actually available* in a year of good precipitation. This unrealistic approach to water rights explains many of the state's water supply issues.

Six major systems of aqueducts and associated infrastructure redistribute water within California: the SWP, the Central Valley Project, a number of Colorado River delivery systems, the Los Angeles Aqueduct, the Tuolumne River/Hetch Hetchy system, and the Mokelumne Aqueduct to the East Bay. The state, federal, and regional agencies that operate these transportation systems are, in most cases, wholesalers that pass their life-giving product on to hundreds of local districts for delivery to retail consumers.

THE STATE WATER PROJECT

The Department of Water Resources (DWR) operates the massive California SWP, the largest state-built multipurpose water project in the United States (maps 13, 14). The SWP moves water from the Feather River watershed in the Sacramento Valley to urban and industrial consumers (70 percent of its contracts) and the balance to agricultural irrigation districts (mostly in Kern County in the San Joaquin Valley). More than two-thirds of Californians receive some of their water from the SWP. About 2.3 MAF are delivered in average years, although the overcommitted system has contracted to deliver 4.2 MAF.

One of the surreal circumstances shaping California water policy is that the SWP cannot actually deliver the amounts in its contracts. The true "safe yield" of the SWP is not the current allocation total of 4.17 MAF of water a year. The average amount of water actually delivered between 2000 and 2014 was just 55 percent of that total. This has created "paper water," as opposed to the essence of real life, "wet" water. Paper water must not be used as the supply basis for authorizing new developments, nor marketed in water transfers. Paper water, when viewed as a legal entitlement, drives bad government planning and policies by forcing the real world to accommodate to hydrological wishful thinking. During droughts, news stories often refer to the difference between "entitlements" or "allocations," without recognizing the reality that entitlements have been grossly overpromised in SWP contracts (table 2).

Twenty-nine agencies hold contracts for SWP water. The contractors cover the SWP's major operating costs and have

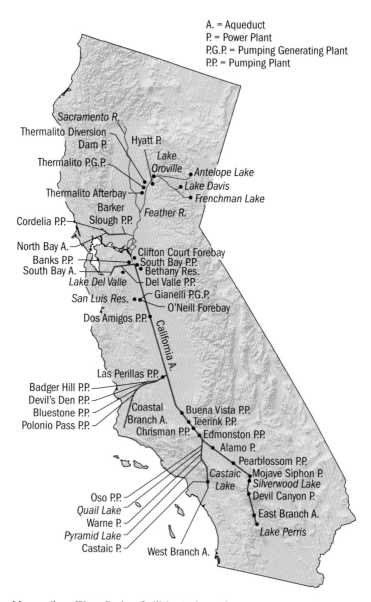

A. = Aqueduct
P. = Power Plant
P.G.P. = Pumping Generating Plant
P.P. = Pumping Plant

Sacramento R.
Thermalito Diversion
Dam P.
Hyatt P.
Thermalito P.G.P.
Lake
Oroville
Antelope Lake
Lake Davis
Thermalito Afterbay
Frenchman Lake
Barker
Slough P.P.
Feather R.
Cordelia P.P.
North Bay A.
Clifton Court Forebay
Banks P.P.
South Bay P.P.
South Bay A.
Bethany Res.
Lake Del Valle
Del Valle P.P.
Gianelli P.G.P.
San Luis Res.
O'Neill Forebay
Dos Amigos P.P.
California A.
Las Perillas P.P.
Badger Hill P.P.
Devil's Den P.P.
Buena Vista P.P.
Coastal
Bluestone P.P.
Branch A.
Teerink P.P.
Polonio Pass P.P.
Chrisman P.P.
Edmonston P.P.
Alamo P.
Pearblossom P.P.
Castaic
Mojave Siphon P.
Lake
Silverwood Lake
Oso P.P.
Devil Canyon P.
Quail Lake
Warne P.
East Branch A.
Pyramid Lake
Lake Perris
Castaic P.
West Branch A.

Map 13. State Water Project facilities (redrawn from two maps in California Department of Water Resources 1998).

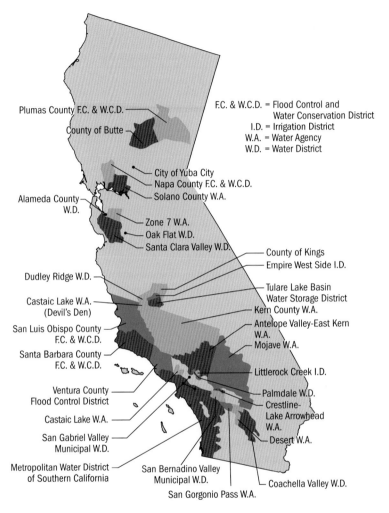

Plumas County F.C. & W.C.D.

County of Butte

F.C. & W.C.D. = Flood Control and
             Water Conservation District
I.D. = Irrigation District
W.A. = Water Agency
W.D. = Water District

City of Yuba City
Napa County F.C. & W.C.D.
Solano County W.A.

Alameda County
W.D.

Zone 7 W.A.
Oak Flat W.D.
Santa Clara Valley W.D.

County of Kings
Empire West Side I.D.

Dudley Ridge W.D.

Tulare Lake Basin
Water Storage District
Kern County W.A.

Castaic Lake W.A.
(Devil's Den)

Antelope Valley-East Kern
W.A.

San Luis Obispo County
F.C. & W.C.D.

Mojave W.A.

Santa Barbara County
F.C. & W.C.D.

Littlerock Creek I.D.

Ventura County
Flood Control District

Palmdale W.D.
Crestline-
Lake Arrowhead
W.A.

Castaic Lake W.A.

Desert W.A.

San Gabriel Valley
Municipal W.D.

Metropolitan Water District
of Southern California

San Bernadino Valley
Municipal W.D.

Coachella Valley W.D.

San Gorgonio Pass W.A.

Map 14. State Water Project service areas (redrawn from Hundley Jr. 2001).

TABLE 2

State Water Project allocations (as a
percentage of entitlement requests)

Actual allocations as percentage of requests (average 2000–2014: 59 percent)	
2000	90
2001	39
2002	45
2003	90
2004	65
2005	90
2006	100
2007	60
2008	35
2009	40
2010	50
2011	80
2012	65
2013	35
2014	5

chipped away very slowly at the $1.75 billion bond debt that funded the initial construction. Since 1960, the SWP has built 29 dams, 18 pumping plants, five hydroelectric power plants, and about 600 miles of canals and pipelines. Four additional combination pumping/generating plants move water uphill into storage basins when electricity costs are low (off-peak hours), then generate power by releasing the same water through turbines during peak energy demand periods.

The SWP system begins 600 miles north of its southern-most service area with the Lake Davis, Frenchman Lake, and Antelope Lake reservoirs on upper tributaries of the Feather River (fig. 50). Oroville Dam, where the Feather River passes out of the foothills,

Figure 50. Antelope Dam and reservoir, in the headwaters region of the Feather River and the SWP.

forms the largest SWP reservoir (fig. 51). The dam towers 770 feet above the riverbed; it is the tallest in the United States. When full, its reservoir covers 15,000 acres with 165 miles of shoreline, and holds 3.5 MAF (2.7 MAF for water supply, 800,000 AF for flood control). A power plant is located *underneath* the reservoir to maximize hydroelectric generation. It is an eerie feeling to visit that plant and know you are standing beneath hundreds of feet of water.

SWP water travels from the Lake Oroville reservoir along the natural channel of the Feather River and enters the Sacramento River. At the Sacramento–San Joaquin Delta, some is pumped into the North Bay Aqueduct toward Napa and Solano Counties. More is diverted by powerful pumps at the Harvey O. Banks Delta Pumping Plant, which pull the SWP's allotment of Delta water into the Bethany Reservoir and the start of the California Aqueduct. In an average year, about 2.1 MAF (of the 2.3 MAF total supplied by the SWP) are extracted here from the Delta. The 444-mile-long "Governor Edmund G. Brown California Aqueduct," to use its formal name, extends southward along the west side of the San Joaquin Valley as a concrete-lined, open canal (fig. 52). Bethany Reservoir is also the point where water is pumped into a side branch, the South Bay Aqueduct, which heads for Alameda and Santa Clara Counties.

Sixty-three miles to the south is San Luis Reservoir, off Highway 152 below Pacheco Pass in the Diablo Range. Water is pumped uphill into this off-stream storage facility. The reservoir, its forebay, and its pumping plants are jointly operated by the federal Central Valley Project (CVP) and DWR. CVP water moves from the Delta to this location in a separate, parallel aqueduct, the Delta–Mendota Canal. Southward, for 103 miles, the aqueduct carries both "kinds" of water.

The California Aqueduct parallels Interstate 5, passing beneath that other important transportation corridor several times (fig. 53).

Figure 51. Oroville, the primary SWP storage reservoir for Feather River water.

Figure 52. The California Aqueduct south of the Delta.

Figure 53. Dos Amigos Pumping Plant on the California Aqueduct, which parallels Interstate 5 in the Central Valley.

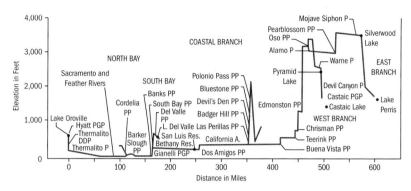

Figure 54. Elevation changes of water moving in the State Water Project.

The Coastal Branch Aqueduct splits away, 185 miles south of the Delta, to direct some water toward the coastal cities of San Luis Obispo, Santa Maria, and Santa Barbara. Central Coast voters decided to fund this artery during the drought that ended in 1993.

The balance of SWP water flows to the south end of the San Joaquin Valley, where it encounters a 2,000-foot-tall barrier, the Tehachapi Mountains. Here the A.D. Edmonston Pumping Plant lifts the water 1,926 feet into 10 miles of tunnels and siphons that pass through the mountains (figs. 54, 55). The SWP is California's largest energy consumer, and this pumping plant burns more energy than any other single user or facility in the state. Although the hydroelectric generating plants of the SWP can, together, generate 5.8 billion kilowatt-hours per year (2.2 billion kilowatt-hours at the Oroville reservoir alone), that is only three-fourths of the electricity consumed in lifting SWP water uphill through the Central Valley and making the massive lift over the Tehacha-pis. Water is heavy! Carry an eight-pound gallon bucket or jug the next time you go upstairs to better appreciate that fact. Each

Figure 55. The A.D. Edmonston Pumping Plant, which lifts SWP water almost 2,000 feet over the Tehachapi Mountains.

Figure 56. Castaic Pumping Plant on the west branch of the California Aqueduct, after it enters Southern California.

AF demands 3,000 kilowatt-hours of electricity to overcome the Tehachapis and be redistributed to the Southland.

Below the Tehachapis, the California Aqueduct divides. The west branch stores water in the Castaic and Pyramid Lake reservoirs in north Los Angeles County to serve coastal cities (fig. 56). The east branch flows by the Mojave Desert city of Palmdale, in the Antelope Valley, and stores water in the Silverwood Lake reservoir in the San Bernardino Mountains (figs. 57, 58). The final SWP facility in this long chain of transportation infrastructure is Lake Perris, a reservoir in Riverside County (fig. 59). The Metropolitan Water District of Southern California (MWD) is

Figure 57. California Aqueduct and new housing made possible by its water deliveries to the Mojave Desert.

Figure 58. Silverwood, an SWP reservoir in the San Bernardino Mountains.

Figure 59. Windsurfer on Lake Perris, the terminal reservoir on the east branch of the California Aqueduct.

the largest SWP contractor, taking more than 2 MAF each year (48 percent of the SWP's contracted water).

The DWR operates three SWP visitor centers for the public: Lake Oroville Visitors Center, Romero Visitors Center at San Luis Reservoir, and Vista del Lago at Pyramid Lake.

THE CENTRAL VALLEY PROJECT

The CVP was conceived to tame seasonal flooding and to shift water southward to irrigate three million acres of drier farmlands (15 percent of CVP water goes to urban/industrial uses). President Franklin D. Roosevelt signed the measure that authorized the CVP and transformed the Central Valley into one of the most important agricultural regions on Earth. The CVP is operated by the U.S. Bureau of Reclamation (although some of the facilities were built by the Army Corps of Engineers). One of the largest water systems in the world, it stores seven MAF, or about 17 percent of the state's developed water, and delivers it to 139 landowners and eight water districts (map 15).

Broader in scope and scale than the SWP (which primarily draws on the Feather River watershed), the CVP dams and diverts water from five major rivers: the Trinity (Trinity Dam), the Sacramento (Shasta Dam), the American (Folsom Dam), the Stanislaus (New Melones Dam), and the San Joaquin (Friant Dam). One of its original goals was to end groundwater overdrafting in the southern half of the Central Valley. Instead, more acreage went into production after CVP water became available, and groundwater pumping actually increased.

Friant Dam, on the San Joaquin River, was completed in 1944, forming Millerton Lake. This was the first of 20 reservoirs in the CVP, which also includes 11 power plants and three fish hatcheries. Shasta Dam spanned the Sacramento River in 1945 (fig. 60). The canal system to deliver irrigation water from Shasta to the San Joaquin Valley opened in 1951. The CVP also dammed the Trinity River and, in 1963, began shipping that water out of the North Coast region to Whiskeytown Reservoir, which passes it along to Shasta.

Clair Engle Lake

Goose Lake

Trinity R.

Keswick Dam

Shasta Lake

Whiskeytown Lake

Shasta Dam

Eagle Lake

Sacramento R.

Tehama-Colusa Canal

Clear Lake

Folsom Dam

Lake Tahoe

Folsom Lake

Sacramento

American R.

Delta Cross Channel

San Francisco

Stanislaus R.

Mono Lake

Contra Costa Canal

San Joaquin R.

Madera Canal

New Melones Dam

Delta–Medota Canal

Millerton Lake

Friant Dam

Friant-Kern Canal

Lake Isabella

Bakersfield

Kern R.

Los Angeles

Salton Sea

San Diego

Map 15. Central Valley Project facilities (redrawn from Hundley Jr. 2001).

Figure 60. Shasta Dam and Reservoir with Mt. Shasta in the distance.

Keswick Dam, nine miles below Shasta, evens out the flow from variable releases through the upstream power plants. Fish-trapping facilities there catch salmon and move them into the nearby Coleman Fish Hatchery, operated by the U.S. Fish and Wildlife Service.

Red Bluff Diversion Dam raises the Sacramento River 17 feet, creating a gravity "head" so water will leave the river and enter the Tehama-Colusa and Corning Canals. One-third of the river's water is diverted into those canals to irrigate 300,000 acres of farmland and provide seasonal water to several national wildlife refuges along the west side of the Sacramento Valley. Red Bluff Dam's fish ladders were never effective and the dam became a major barrier to migrating fish. After years of court fighting, the dam's gates are now kept open eight months of the year, and the intake to the canals has been reengineered with improved fish screens.

Water stored behind Shasta Dam can be moved 450 miles to Bakersfield, traveling the first leg of that journey in the Sacra-

Figure 61. Pumping from the Delta into the Delta–Mendota Canal, part of the CVP.

mento River itself. At Sacramento, American River water stored behind Folsom Dam is added. About 2.5 MAF are pumped annually from the Delta at the Tracy Pumping Plant (not far from the SWP pumps that serve the California Aqueduct) into the Delta–Mendota Canal (fig. 61). Some CVP water is stored in San Luis Reservoir and, from there, shares space within the California Aqueduct, which parallels the Delta–Mendota Canal but heads farther south.

New Melones Dam on the Stanislaus River was completed in 1979. Before the gates were closed to impound the Stanislaus, a fight developed, because the reservoir would flood a spectacular river canyon that, among other values, was popular for whitewater rafting. New Melones Reservoir was filled in 1982.

Just below the confluence of the north and middle forks of the American River, construction of the Auburn Dam began in 1974, but it was halted the next year because of seismic risks. Plans for that dam, in a number of manifestations, seemed a "never-ending story" in California's perennial water debates, until the project's water rights were finally canceled in 2008.

Operation of the CVP has generated controversies about environmental degradation, the prices charged (or subsidies given) to farmers, and lax enforcement of acreage limitations. Bureau of Reclamation water was meant to serve farms limited to 160 acres, to encourage small farmers who lived on their land. That type of land ownership was never the broad pattern in California, however. Under Spanish and Mexican land grants, just a few individuals owned large ranches. Federal land grants later transferred 11 percent of California's acreage to railroads, with much of that land spanning the Central Valley. A few entrepreneurs manipulated federal homestead, timber, and swampland programs to circumvent acreage limits and acquire massive parcels. Henry Miller and Charles Lux ultimately controlled about 750,000 acres along both sides of the San Joaquin River for 100 miles, plus a 50-mile stretch along the Kern River. Similarly, James Ben Ali Haggin acquired more than 400,000 acres in the Central Valley. A transition from a few large landholdings to many smaller farms was conceivable, but the vision of many thousands of farms limited to 160 acres was never actually realized.

In 1982, a reform act increased CVP acreage limitations to 960 acres and dropped the former residency requirement. Today, however, 80 percent of the huge farms still exceed 1,000 acres.

Another major reform came in 1992, when the Central Valley Project Improvement Act (CVPIA) elevated fish and wildlife protection and restoration to primary purposes of the CVP. To correct environmental damage, including endangered species

Figure 62. The Friant-Kern Canal, heading southward below Friant Dam and leaving very little water for the San Joaquin River, seen in the distance.

designations for native fish, 800,000 AF of annual runoff were dedicated back to the environment (600,000 AF during dry years). Also, CVP contractors were allowed to market water to buyers outside the CVP service area.

One of the biggest environmental consequences of the CVP was not corrected. At Friant Dam, the entire flow of the San Joaquin River had been diverted into the Friant-Kern and Madera Canals. The Friant-Kern Canal travels southward 152 miles from Friant Dam, and the Madera Canal extends 36 miles northward. A little water was allowed down the San Joaquin River channel for 28 miles, to Gravelly Ford, to serve users holding priority riparian rights, but from there the channel went dry until Mendota Pool (fig. 62).

Mendota Pool is a strange place in California's redesigned waterscape. There the Delta–Mendota Canal delivers Sacramento

Figure 63. Mendota Pool, where Sacramento River water taken from the Delta is delivered to the dry San Joaquin River channel to serve CVP exchange contractors.

River water to serve CVP "exchange contractors." Irrigators in the San Joaquin Valley who held water rights to the San Joaquin River's natural flow agreed to accept this alternate CVP water, provided as the highest priority right coming out of the Delta. They take delivery of their exchange water at Mendota Pool (fig. 63) and soon divert it to farm fields. The San Joaquin channel dries up again not far below Mendota. Other than in very wet years, the San Joaquin River actually was *not* a river for about 60 miles, until the Merced entered its channel with a transfusion from the Yosemite watershed. The Tuolumne and Stanislaus added water farther north. Some agricultural drainage water returned to the channel, carrying salts, fertilizers, and pesticides, and some treated municipal effluent was returned as well.

The San Joaquin River, before it finds its rather pitiful way to the Delta, has been called "the lower colon of California."

Of course, after the CVP diverted the San Joaquin's water to grow crops, other life that depended on the river diminished or disappeared. In particular, 100,000 salmon could no longer reach spawning grounds. After a consortium led by the Natural Resources Defense Council (NRDC) sued and prevailed in the courts, a settlement agreement between the consortium and the Friant Water Users Authority led to restoration efforts. The agreement adopted two guiding principles: that the river's natural functions be restored and that irrigators not lose water or additional money. In March 2010, flows were returned to the river, reconnecting it to the Delta for the first time since the 1940s. But issues of seepage into adjacent farm fields delayed full implementation of the restoration plan. In 2015, if there is enough water available (after three preceding years of drought), flows will resume. The plan includes spring and fall pulses to benefit salmon, with low flows the rest of the time.

COLORADO RIVER DELIVERY SYSTEMS

The Colorado River is currently the source for 37 percent of Southern California's urban water consumption and 92 percent of the southern counties' farm irrigation water. Six other states along the river's watershed, plus the nation of Mexico, share allocated portions of the river's flow (map 16). California's allotment became 4.4 MAF per year. Three-fourths of that was to irrigate 900,000 acres of farmland. Priority water rights are held by three irrigation districts. The Palo Verde Irrigation District (PVID) is located 110 miles north of the Mexican border, just west of the river. The Imperial Irrigation District (IID), in

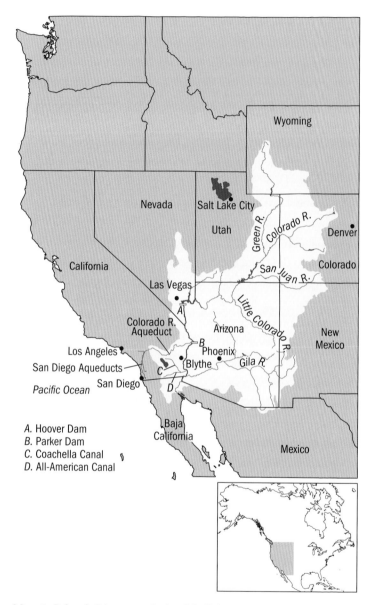

Map 16. Colorado River watershed and facilities (redrawn from Water Education Foundation 1991).

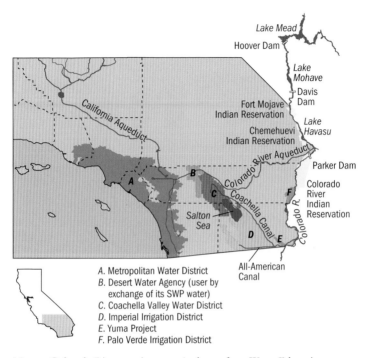

Map 17. Colorado River service areas (redrawn from Water Education Foundation 1991).

Imperial County, south of the Salton Sea, receives the majority of the state's Colorado River allotment via the All-American Canal. The Coachella Valley Water District (CVWD) serves farms north of the Salton Sea, in parts of Riverside, Imperial, and San Diego Counties (map 17).

The fourth priority water right is held by the MWD. Member agencies of "the Met" include 14 cities, 12 municipal water districts, and one county water authority. It is the water wholesaler to 95 percent of the South Coast region. Along that coastal plain, from Ventura County down to the Mexican border, the MWD provides about 60 percent of the water supply,

Figure 64. Parker Dam, which forms Lake Havasu, the diversion point for Colorado River water into the Colorado Aqueduct.

drawing from both the Colorado River and the SWP. The Colorado River Aqueduct moves water 242 miles across the Mojave Desert.

Lake Mead, behind Hoover Dam, is the primary storage reservoir in the lower Colorado River basin. The Colorado River Aqueduct begins at Parker Dam, 155 miles downstream from Hoover. The aqueduct can handle 1.2 MAF annually, or more than 1 billion gallons a day (figs. 64, 65).

The newest large reservoir in California was completed by the MWD in 1999. Diamond Valley Reservoir, near Hemet, was built to alleviate concerns that aqueducts feeding Southern California could be cut off by earthquakes and to provide optional water during droughts. The off-stream reservoir was filled in 2002 with 800,000 AF of water (a six-month supply for the MWD) from the Colorado River Aqueduct and the California Aqueduct.

Figure 65. The Colorado River Aqueduct crossing the desert.

Every drop of Colorado River water has been allocated to water rights holders. In fact, the river is also over-allocated, because apportionments to the states were based on overestimates of the annual runoff. In *Cadillac Desert* (1986, 126), Marc Reisner described the Colorado as "unable to satisfy all the demands on it, so it is referred to as a 'deficit' river, as if the river were somehow at fault for its overuse." Of course, the Colorado is not to blame. The river itself is not a water rights holder.

In a treaty, the United States promised to pass along 1.5 MAF of Colorado River water as far as the border of Mexico. In most years, diversions inside Mexico left nothing to reach the Gulf of Mexico. The wetlands estuary, vibrant with life at the mouth of the Colorado, died of thirst. However, "Minute 319," was signed on November 20, 2012, to modify the 1944 treaty. Under this historic agreement, 158,088 AF of river water will again flow all the way to the Colorado Delta. That total, to be delivered over a period of five years, is only about one percent of the river's historic flow, but within days, once water began reaching the estuary, life was seen returning to the wetlands that had been dried by diversions during the twentieth century.

For many years, California took more than its allocated 4.4 MAF of Colorado River water, because other states in the lower river basin were not prepared to divert their full allotments. Southern Californians became accustomed to about 800,000 AF of surplus to keep the MWD's Colorado River Aqueduct full; they took a total of 5.3 MAF in 2000. But by then, Arizona had plumbing in place to handle all of its allotment, and Nevada, with the Las Vegas region booming, was asking to exceed its portion.

The Secretary of the Interior controls use of the river's "surplus" water. Under pressure from the other Colorado River states, the secretary ordered California to show good progress toward weaning itself from the extra 800,000 AF or face mandatory cuts. The Colorado River Water Use Plan, or "4.4 Plan," was to be ready by December 31, 2002, and had to convince the other watershed states that it was realistic. Planners aimed to reallocate 800,000 AF annually within Southern California without pulling more from northern California's already heavily impacted rivers. That meant that Imperial and Coachella Valley agriculture had to give up water. The alternatives of urban demand reduction or population stabilization to live within the limits of the existing water supply were not given serious consideration.

A list prepared for the 4.4 Plan included a transfer of 110,000 AF between the MWD and the IID, a transfer of 200,000 AF between the IID and the San Diego County Water Authority (SDCWA), a shift of 100,000 AF between the IID and the

CVWD, the lining of irrigation canals with concrete to control seepage and save 94,000 AF, and groundwater sources totaling 300,000 AF in Arizona, the Mojave Desert, and the Coachella Valley.

There were problems with almost all of the proposals. Groundwater banking for future withdrawals was not as controversial as pumping of native groundwater. Conservation measures, such as concrete lining of irrigation canals, were not as controversial as land fallowing, which would shut down extended economies in farm communities. In the Imperial Valley, fallowing would also exacerbate the salinity issues of the Salton Sea, because irrigation flows to the sea would be cut back. The IID balked at land fallowing and wanted assurances that it would not be liable for damages to the Salton Sea environment.

When the deadline arrived with no agreement, the Department of the Interior announced it would immediately reduce the MWD's access to surplus water by about 415,000 AF. It also punished the IID by shifting about 200,000 AF from the farmers to the MWD (though the state's overuse had all originated with the urban wholesaler). The IID vowed to legally fight the attack on its historic water rights, yet the forced transfer of agriculture water to serve urban interests seemed to be a sign of California's future.

Back in the year 2000, Lake Mead was nearly full. The following 14 years were the driest in the Colorado Basin in the last century. The shared demand for Colorado River water exceeded supply by about 30 percent during the drought, leaving the massive reservoirs, Mead and Powell, extremely low. Groundwater pumping has been relied on to make up the difference between thirst and supply.

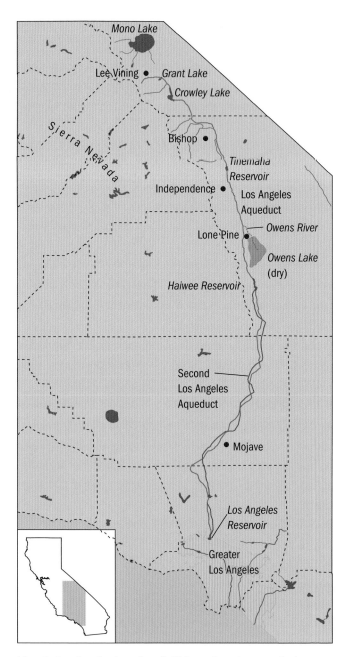

Map 18. Los Angeles Aqueduct facilities and service areas (redrawn from Los Angeles Department of Water and Power 1988).

Figure 66. Lee Vining Creek at the point where water is diverted from the Mono Lake tributary into the Los Angeles Aqueduct.

THE LOS ANGELES AQUEDUCT

Completed in 1913, the Los Angeles Aqueduct brings water from the Eastern Sierra to the city of Los Angeles (map 18). Construction of this aqueduct and its associated reservoirs was the first major long-range water delivery project in California. The city acquired water rights by purchasing 300,000 acres of the Owens Valley, or about 98 percent of all the private land in that Eastern Sierra valley, hundreds of miles north of Los Angeles. The imperialistic pressure exerted by a distant city on unwilling rural landowners became an inglorious part of California's water history. The aqueduct was extended into the Mono Basin and began diverting streams away from Mono Lake in 1941 (figs. 66, 67).

Figure 67. Los Angeles Aqueduct pipe in the upper Owens Valley.

Figure 68. The cascade, visible from the Golden State Freeway,
where Los Angeles Aqueduct water enters the San Fernando Valley.

The Los Angeles Department of Water and Power moves an average of 400,000 AF of Eastern Sierra water to the city each year—enough to serve about 3.2 million people. With additional MWD water and some local groundwater, the Los Angeles population has grown to four million, at least eight times the number that local water supplies would have allowed (fig. 68). That growth is one of the clearest examples of William Mulholland's observation "Whoever brings the water, brings the people." Mulholland was the engineer who oversaw the early design, construction, and operation of the Los Angeles Aqueduct system.

After four decades of stream diversions from the Mono Lake basin, damage to the lake and the dewatering of its tributary streams fostered an environmental battle in the 1980s and victory for the lake defenders in 1994. Other issues arose when dust from the bed of Owens Lake, which was completely dried up by the diversions, became a major air pollution source at the south end of the Owens Valley. To stabilize Mono Lake, correct violations of air quality laws, and rewater portions of the lower Owens River, Los Angeles has found ways to reduce its reliance on Eastern Sierra water. Most of that reduction has been achieved through water conservation. The city pursued an aggressive program of toilet replacement, offering free low-flush toilets to its customers. This and other conservation and water-recycling efforts allowed Los Angeles to grow by 30 percent during the final decades of the twentieth century, yet see a seven percent *decrease* in its total water use and a 15 percent drop in per capita demand. In October 2014, the Mayor of Los Angeles issued an executive directive to further reduce water use by 20 percent by the year 2017 and cut imported water by 50 percent by 2024, through reliance on even more conservation, by cleaning

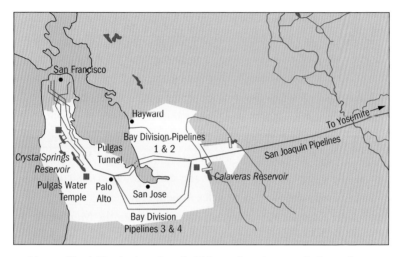

Map 19. Hetch Hetchy Aqueduct, facilities, and service areas (redrawn from San Francisco Public Utilities Commission, n.d.).

local groundwater that had become polluted, and by recycling wastewater.

THE HETCH HETCHY AQUEDUCT

The Hetch Hetchy Aqueduct brings Tuolumne River water to 2.3 million people in San Francisco and other portions of the Bay Area (map 19). The system originates at Hetch Hetchy Valley, inside Yosemite National Park, where O'Shaughnessy Dam was completed in 1923 to dam the Tuolumne River (figs. 69, 70). The water system of the San Francisco Public Utilities Commission (SFPUC) also includes five reservoirs in the Bay Area: two in Alameda County (San Antonio and Calaveras) and three on the Peninsula south of San Francisco (San Andreas, Crystal Springs, and Pilarcitos). Those reservoirs supplement the system with

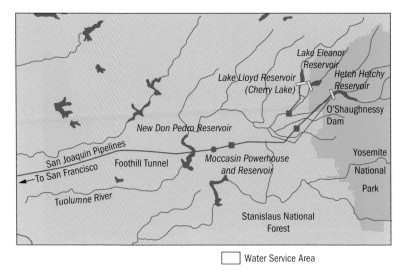

Map 19. *(continued)*

runoff from local watersheds, but 85 percent of the city's water comes from the Tuolumne River. About two-thirds of that Hetch Hetchy water is sold to 26 other cities and water districts in San Mateo, Santa Clara, and Alameda Counties.

The system also generates electricity—a major revenue source for San Francisco (fig. 71). After water leaves Hetch Hetchy, it passes through tunnels leading down to a series of powerhouses. Three massive pipes then bring water across the Central Valley. The aqueduct that delivers water to San Francisco is entirely gravity fed, so a 25-mile-long tunnel had to be constructed through the Coast Ranges. One segment of the aqueduct crosses beneath the south part of the bay and another portion swings south of the bay. They rejoin at the Pulgas Water Temple, a monument marking the arrival point for aqueduct water on the Peninsula.

Figure 69. Hetch Hetchy Valley, 1913.

Figure 70. O'Shaughnessy Dam and Hetch Hetchy Reservoir, inside Yosemite National Park.

Figure 71. Moccasin, a hydroelectric power plant that is part of the SFPUC's Hetch Hetchy system.

Some of the exceptionally clean Tuolumne water goes directly into the city's water system, and some is stored in the Peninsula reservoirs. Hetch Hetchy water that goes directly to city pipelines does not have to be filtered, because it exceeds water quality standards (it is disinfected with chlorine). Only a few cities in the nation are authorized to use unfiltered water.

Hetch Hetchy Valley, where that clean water is first stored, is about two-thirds the size of, and very similar in character to, the more famous Yosemite Valley. The Tuolumne River meandered through its meadows and trees, beneath towering granite cliffs and domes punctuated by waterfalls. The idea of building a dam inside a national park, and in that particular valley, infuriated John Muir. "These temple destroyers, devotees of ravaging commercialism, seem to have a perfect contempt for Nature, and, instead of lifting their eyes to the God of the mountains, lift them to the Almighty Dollar," Muir declared in 1912, as congressional hearings were about to start to consider the project. "Dam Hetch Hetchy? As well dam for water tanks the people's cathedrals and churches, for no holier temple has ever been consecrated by the heart of man" (1912, 181).

Muir argued that there were other rivers available to San Francisco, including the Mokelumne (which would be developed soon after for East Bay cities). Congress authorized the reservoir construction by passing the Raker Act in 1913. When President Woodrow Wilson signed the law, he stated that domestic water supply was the "highest use" of the river water, even though it originated in a national park. But much of the nation was appalled, and two years after authorizing the dam, Congress passed the National Parks Act, ensuring that dam construction would never again be allowed within federal parks.

Concerns about the old, deteriorating Hetch Hetchy Aqueduct system's ability to withstand earthquakes led to a $1.7 billion bond, approved by San Francisco voters in the November 2002 election, to fund repairs and upgrades. Because two-thirds of the SFPUC's water customers had no say in its management decisions, some of their contracting agencies had also pushed for state legislation to require oversight of the repairs, and even for a new regional utility district something like the MWD to oversee water distributions.

Those who yearn to see the restoration of Hetch Hetchy Valley envision removal of O'Shaughnessy Dam from Yosemite National Park. The Restore Hetch Hetchy organization proposes developing alternative water storage downstream on the Tuolumne River, with no loss of water for San Francisco. In 2012, San Francisco voters defeated a ballot measure that would have required the City to study the feasibility of removing the dam. Despite that setback, the Restore Hetch Hetchy organization continues to push for water conservation and alternative supply alternatives to improve Bay Area water reliability and make that long-term dream possible.

THE MOKELUMNE AQUEDUCT

"East Bay MUD" is the nickname for the East Bay Municipal Utility District, which serves 35 communities in Alameda and Contra Costa Counties, including Berkeley and Oakland, and parts of the San Ramon Valley (map 20). It is an earthy nickname for an agency that supplies domestic water to about 1.2 million people.

The Mokelumne River in the central Sierra Nevada is the source for virtually all of EBMUD's water. The watershed drains parts of Alpine, Amador, and Calaveras Counties. The utility built the Pardee Dam across the Mokelumne in the foothills northeast of Stockton. Pardee can hold a 10-month supply of water. Below Pardee is Camanche Reservoir, which helps regulate releases to serve downstream water rights holders and the fisheries and riparian habitat needs of the lower river.

Almost 30,000 acres in the Mokelumne River watershed belong to EBMUD and 25,000 acres of other watershed lands in the East Bay. After a 30- to 45-hour trip, water in the system has traveled 91.5 miles across the Central Valley via the Mokelumne and Lafayette Aqueducts to enter East Bay reservoirs or filter plants. Basins in the Berkeley and Oakland Hills contain the San Pablo, Briones, Lafayette, Upper San Leandro, and Chabot Reservoirs (fig. 72).

EBMUD also holds an American River entitlement that could be sent to the Mokelumne Aqueduct via the Folsom South Canal. This supplemental supply has only been tapped once,

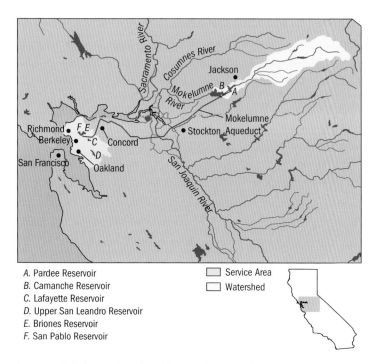

A. Pardee Reservoir
B. Camanche Reservoir
C. Lafayette Reservoir
D. Upper San Leandro Reservoir
E. Briones Reservoir
F. San Pablo Reservoir

□ Service Area
□ Watershed

Map 20. Mokelumne Aqueduct/East Bay MUD facilities and service areas (redrawn from Montgomery 1999).

during the drought year of 1977 to 1978, when it was actually pumped from the Delta. This intake point generated controversy. EBMUD preferred the cleaner water taken from the American River, but environmentalists and the city of Sacramento were concerned about the impacts of such diversions on the river. A decades-long legal battle led to affirmation of EBMUD's water right, but modifications have been negotiated. The intake point was shifted to just downstream from Sacramento, to maintain minimum flows in the American until it merges with the Sacramento River.

Figure 72. EBMUD's Briones Reservoir, east of Tilden Regional Park.

THE NORTH BAY

Some cities north of San Francisco Bay, including Santa Rosa, Petaluma, San Rafael, and Mill Valley, are served by the Sonoma County Water Agency. Their primary water source is the Russian River, via the Santa Rosa, Sonoma, Petaluma, and North Marin Aqueducts and the Cotati Intertie. The agency also taps into the headwaters of the Eel River, diverting that water into the Russian River through a tunnel. The fast-growing cities of Vallejo, Fairfield, and Vacaville are served by the Solano County Water Agency, which captures Putah Creek water in Lake Berryessa and ships it south in the Putah South Canal. SWP water delivered by the North Bay Aqueduct adds to the Solano supply.

The state has thousands of water agencies and districts; 280 retail water agencies serve about 90 percent of California's users. With so many agencies and layers of decision making, coordination (or the lack of it) can be an obstacle to wise management of the resource. The SDCWA is an example of one of the largest regional layers in the water supply hierarchy. SDCWA is an MWD member that imports 84 percent of the county's water. It serves six cities, four rural water districts, three irrigation districts, eight municipal water districts, one public utility district, and a federal military base. The Colorado River provides 73 percent of its water; 27 percent comes from the SWP (both sources are delivered, of course, by the MWD). In addition to SDCWA supplies, the county also relies on local sources (including surface water, groundwater, recycled water, and some water produced by desalination).

Regional decision making has been improved where cities and agencies formed other Joint Powers Water Authorities. The Sacramento Regional Water Authority was established in 2001 by 21 agencies in the Sacramento region. The Stanislaus Regional Water Authority is a more recent example, formed in 2011, by the cities of Ceres, Modesto, and Turlock. And, in 2012, six cities established the Monterey Peninsula Authority, to find ways to meet a State-ordered 2017 deadline to reduce overdrafting from the Carmel River (see the "Integrated Water Management" Section, Chapter 5).

·   ·   ·

Because water is essential for life, every natural habitat and everything that lives in California has been affected by the redistribution of the state's water (see table 3 for one way of looking at this redistribution). The aqueduct systems have fostered economic development and population growth. They have also generated a long list of challenges for Californians to face as they shape their future.

TABLE 3

Where Does Your Water Come From?

Sources: **1.** *Groundwater;* **2.** *Feather River/California Aqueduct/State Water Project;* **3.** *Colorado River/Metropolitan Water District of Southern California*;* **4.** *Owens and Mono Basins/Los Angeles Aqueduct/L.A. Department of Water & Power;* **5.** *Tuolumne River/San Francisco PUC;* **6.** *Mokelumne River/East Bay Municipal Utility District;* **7.** *Russian and Eel Rivers/Sonoma County Water Agency;* **8.** *Lake Berryessa/Putah Creek/Solano County Water Agency;* **9.** *Central Valley Project (many Northern California rivers)/U.S. Bureau of Reclamation;* **10.** *Local reservoirs/streams;* **11.** *Ocean desalination*

Adelanto (1)	Bell Gardens (1), (2), (3)
Agoura Hills (2), (3)	Bellflower (1), (2), (3)
Alameda (6), (10)	Belmont (5), (10)
Albany (5), (10)	Benicia (2), (10)
Alhambra (1), (2), (3)	Berkeley (6), (10)
Aliso Viejo (3)	Beverly Hills (1), (2), (3)
American Canyon (2)	Blythe (1)
Anaheim (1), (2), (3)	Brawley (3)
Antioch (9), (10)	Brea (1), (2), (3)
Apple Valley (1),†	Brentwood (1), (2)
Arcadia (1), (2), (3)	Buena Park (1), (2), (3)
Arcata (1), (10)	Burbank (1), (2), (3)
Arroyo Grande (1), (10)	Burlingame (5)
Artesia (1), (2), (3)	Calabasas (2)
Arvin (1)	Calexico (3)
Atascadero (1)	California City (1), (2)
Atwater (1)	Camarillo (1), (2)
Auburn (10)	Campbell (1), (2), (9), (10)
Avenal (9)	Canyon Lake (1), (2), (3), (10)
Azusa (1), (2), (10)	Capitola (1), (10)
Bakersfield (1), (2), (10)	Carlsbad (2), (3),
Baldwin Park (1), (2), (3)	Carpinteria (1), (2), (10)
Banning (1)	Carson (1), (2), (3)
Barstow (1)	Cathedral City (1)
Beaumont (1)	Ceres (1)
Bell (1), (2), (3)	Cerritos (1), (2), (3)

(continued)

TABLE 3

(continued)

Chico (1)	Duarte (1)
Chino (1), (2)	Dublin (1), (2), (10)
Chino Hills (1), (2)	East Palo Alto (5), (10)
Chowchilla (1)	El Cajon (2), (3), (10)
Chula Vista (1), (2), (3), (10)	El Centro (3)
Citrus Heights (1), (9)	El Cerrito (6), (10)
Claremont (1), (2), (3)	El Monte (1), (2), (3)
Clayton (9), (10)	El Segundo (2), (3)
Clearlake (10)	Elk Grove (1), (10)
Clovis (1), (10)	Encinitas (2), (3), (10)
Coachella (1)	Escondido (2), (3), (10)
Coalinga (9)	Eureka (10)
Colton (1)†	Fairfield (2), (8)
Commerce (1), (2), (3)	Fillmore (1)
Compton (1), (2), (3)	Folsom (9)
Concord (9), (10)	Fontana (1), (2), (10)
Corcoran (1)	Fortuna (1)
Corona (1), (2), (3),	Foster City (5)
Coronado (2), (3), (10)	Fountain Valley (1), (2), (3)
Costa Mesa (1), (2), (3)	Fremont (1), (2), (5), (10)
Covina (1), (2), (3), (10)	Fresno (1), (9), (10)
Cudahy (1), (2), (3)	Fullerton (1), (2), (3)
Culver City (2), (3)	Galt (1)
Cupertino (1), (2), (5), (9), (10)	Garden Grove (1), (2), (3)
Cypress (1), (2), (3)	Gardena (1), (2), (3)
Daly City (1), (5), (10)	Gilroy (1)
Dana Point (2), (3)	Glendale (1), (2), (3)
Danville (6), (10)	Glendora (1), (2), (3)
Davis (1) (UC Davis 2, 8)	Goleta (1), (2), (10)
Delano (1)	Grand Terrace (1)†
Desert Hot Springs (1)	Grass Valley (10)
Diamond Bar (2), (3)	Greenfield (1)
Dinuba (1)	Grover Beach (1), (10)
Dixon (1)	Half Moon Bay (1), (5), (10)
Downey (1)	Hanford (1)

Hawaiian Gardens (1), (3), (10)
Hawthorne (1), (2), (3)
Hayward (5), (6), (10)
Healdsburg (10)
Hemet (1)
Hercules (6), (10)
Hermosa Beach (1), (2), (3)
Hesperia (1)†
Highland (1), (2), (10)
Hillsborough (5), (10)
Hollister (1)
Huntington Beach (1), (2), (3)
Huntington Park (1), (2), (3)
Imperial Beach (2), (3), (10)
Indio (1)
Inglewood (1), (2), (3)
Irvine (1), (2), (3)
King City (1)
Kingsburg (1)
La Canada Flintridge (1), (2), (3), (10)
La Habra (1), (2), (3)
La Mesa (2), (3), (10)
La Mirada (1), (2), (3)
La Palma (1), (2), (3)
La Puente (1), (2), (3)
La Quinta (1)
La Verne (1), (2), (3)
Lafayette (6), (10)
Laguna Beach (2), (3)
Laguna Hills (2), (3)
Laguna Niguel (2), (3)
Laguna Woods (2), (3)
Lake Elsinore (1), (2), (3), (10)
Lake Forest (1), (2), (3)
Lakewood (1), (2), (3)
Lancaster (1), (2)
Larkspur (7), (10)

Lathrop (1), (10)
Lawndale (1), (2), (3)
Lemon Grove (2), (3), (10)
Lemoore (1)
Lincoln (1), (9), (10)
Lindsay (1), (9)
Livermore (1), (2), (10)
Livingston (1)
Lodi (1)
Loma Linda (1)†
Lomita (1), (2), (3)
Lompoc (1)
Long Beach (1), (2), (3)
Los Alamitos (1), (2), (3)
Los Altos (1), (2), (10)
Los Angeles (1), (2), (3), (4)
Los Banos (1)
Los Gatos (1), (2), (9), (10)
Lynwood (1), (2), (3), (10)
Madera (1)
Malibu (1), (2), (3), (4)
Manhattan Beach (1), (2), (3)
Manteca (1), (10)
Marina (1), (11)
Martinez (9), (10)
Marysville (1)
Maywood (1), (2), (3)
McFarland (1)
Menlo Park (5), (10)
Merced (1)
Millbrae (5), (10)
Mill Valley (7), (10)
Milpitas (2), (5), (9)
Mission Viejo (2), (3)
Modesto (1), (10)
Monrovia (1), (2), (3)
Montclair (1), (2)
Montebello (1), (2), (3)

(continued)

TABLE 3

(continued)

Monterey (1), (10)	Parlier (1)
Monterey Park (1), (2)	Pasadena (1), (2), (3), (10)
Moorpark (1), (2)	Paso Robles (1)
Moraga (1), (6), (10)	Patterson (1)
Moreno Valley (1), (2), (3)	Perris (2), (3)
Morgan Hill (1)	Petaluma (1), (7)
Morro Bay (1), (2)	Pico Rivera (1)
Mountain View (1), (2), (5), (9), (10)	Piedmont (6), (10)
	Pinole (6), (10)
Murrieta (1), (2), (3), (10)	Pittsburg (1), (9)
Napa (2), (10)	Placentia (1), (2), (3)
National City (1), (2), (3), (10)	Placerville (10)
Newark (1), (2), (5), (10)	Pleasant Hill (6), (9), (10)
Newport Beach (1), (2), (3)	Pleasanton (1), (2), (10)
Norco (1)	Pomona (1), (2), (3), (10)
Norwalk (1), (2), (3)	Port Hueneme (1), (2)
Novato (7), (10)	Porterville (1)
Oakdale (1)	Poway (2), (3), (10)
Oakland (6), (10)	Rancho Cordova (1), (10)
Oakley (10)	Rancho Cucamonga (1), (2), (10)
Oceanside (1), (2), (3), (11)	Rancho Mirage (1)
Ontario (1), (2)	Rancho Palos Verdes (1), (2), (3)
Orange (1), (2), (3), (10)	Rancho Santa Margarita (2), (3)
Orinda (6), (10)	Red Bluff (1)
Oroville (1), (2), (10)	Redding (1), (9)
Oxnard (1), (2)	Redlands (1), (2), (10)
Pacific Grove, (1), (10)	Redondo Beach (1), (2), (3)
Pacifica (5)	Redwood City, (1), (10)
Palm Desert (1)	Reedley (1)
Palm Springs (1), (10)†‡	Rialto (1), (10)
Palmdale (1), (2), (10)	Richmond (6), (10)
Palo Alto (1), (5)	Ridgecrest (1)
Palos Verdes Est. (2), (3)	Ripon (1)
Paradise, (10)	Riverbank (1)
Paramount (1), (2), (3)	Riverside (1)

Rocklin (10)
Rohnert Park (1), (7)
Rosemead (1), (2), (3)
Roseville (1), (9)
Sacramento (1), (10)
Salinas (1), (10)
San Anselmo (7), (10)
San Bernardino (1)†, (2), (10)
San Bruno (1), (5), (10)
San Carlos (5)
San Clemente (1), (2), (3)
San Diego (2), (3), (10)
San Dimas (1), (2), (3)
San Fernando (1), (2)
San Francisco (5), (10)
San Gabriel (1), (2), (3)
San Jacinto (1), (2), (3), (10)
San Jose (1), (2), (5), (9), (10)
San Juan Capistrano (1), (2), (3)
San Leandro (6), (10)
San Luis Obispo (1), (10)
San Marcos (2), (3)
San Marino (1), (2), (3)
San Mateo (5), (10)
San Pablo (6), (10)
San Rafael (7), (10)
San Ramon (6), (10)
Sanger (1)
Santa Ana (1), (2), (3)
Santa Barbara (1), (2), (10)
Santa Clara (1), (2), (5), (9), (10)
Santa Clarita (1), (2)
Santa Cruz (1), (10)
Santa Fe Springs (1), (2), (3)
Santa Maria (1), (2), (10)
Santa Monica (1), (2), (3)
Santa Paula (1)
Santa Rosa (1), (7), (10)

Santee (2), (3)
Saratoga (1), (2), (9), (10)
Scotts Valley (1)
Seal Beach (1), (2), (3)
Seaside (1)
Selma (1)
Shafter (1)
Sierra Madre (1), (2), (10)
Signal Hill (1), (2), (3)
Simi Valley (1), (2)
Solana Beach (2), (3), (10)
Soledad (1)
South El Monte (1), (2), (3)
South Gate (1), (2), (3)
South Lake Tahoe (1), (10)
South Pasadena (1), (2), (3)
South San Francisco (1), (5)
Stanton (1), (2), (3)
Stockton (1), (10)
Suisun City (2), (8)
Sunnyvale (1), (2), (5), (9), (10)
Susanville (1), (10)
Tehachapi (1)
Temecula (1), (2), (3), (10)
Temple City (1), (2), (3)
Thousand Oaks (2)
Torrance (1), (2), (3)
Tracy (1), (9)
Truckee (1)
Tulare (1)
Turlock (1)
Tustin (1), (2), (3)
Twentynine Palms (1)
Ukiah (1)
Union City (1), (2), (5)
Upland (1), (2), (10)
Vacaville (1), (2), (8)
Vallejo (2), (8), (10)

(continued)

TABLE 3

(continued)

Ventura (1), (10)	West Sacramento (1), (9), (10)
Victorville (1)†	Westminster (1), (2), (3)
Visalia (1)	Whittier (1)
Vista (1), (2), (3), (10)	Windsor (1), (7)
Walnut (2), (3)	Woodland (1)
Walnut Creek (6), (9), (10)	Yorba Linda (1), (2), (3)
Wasco (1)	Yuba City (1), (2), (10)
Watsonville (1), (10)	Yucaipa (1), (2), (10)
West Covina (1), (2), (3), (10)	Yucca Valley (1)†
West Hollywood, (east) (4); (west) (1), (2), (3)	

* MWD is also the largest customer for SWP water.

† Groundwater is replenished with SWP water.

‡ Palm Springs groundwater is replenished with Colorado River water in exchange for SWP allotment.

SOURCES: Water Education Foundation (www.watereducation.org/where-does-my-water-come) and specific water district sources. Cities listed have populations greater than 10,000 in 2010.

CHAPTER FOUR

Challenges to California Water Management

Historically, water development in California may have had more of an impact on biodiversity than any other single factor.
— Andrew Cohen, "The Hidden Costs of California's Water"

Each of us has had a toxic waste hauler dumping surreptitiously into the groundwater for us by proxy. We may not do this ourselves, but we willingly participate in a system that fosters such destruction.
— Arthur Versluis, "The Waters Under the Earth"

CLIMATE CHANGE AND THE WATER CYCLE

Descriptions of historic precipitation and snowmelt runoff averages have become less meaningful, as the state's climate and weather change toward a new "normal." Since the nineteenth century, atmospheric carbon dioxide levels rose from about 270 parts per million (ppm) to over 400 ppm (fig. 73). The Earth's atmosphere has not held that much carbon dioxide for millions of years, back at a time when the planet was far warmer and sea levels much higher. The rise coincided with the industrial

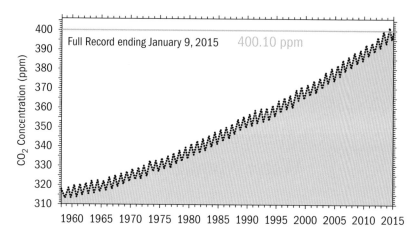

Figure 73. Carbon dioxide in the atmosphere, measured at the Mauna Loa Observatory, 1959 to January 2015.

revolution and accelerated as the Earth's human population grew from 1.6 billion people in 1900 to about 7 billion in 2014. The scale of human activities in California has become so massive that the state, by itself, plays a significant role in global climate change. With a population of 38.3 million people in 2015, the state held 0.5 percent of the Earth's population, yet accounted for two percent of the fossil fuel burned worldwide. In the second decade of the twenty-first century, California was generating over 450 million tons of CO_2 a year, according to the California Energy Commission.

Global warming has increased because carbon dioxide is a "greenhouse gas." Some of the sunlight energy that passes through the air to the Earth's surface radiates as heat back into the atmosphere, where molecules like carbon dioxide, methane, nitrous oxide, ozone, and water vapor absorb the infrared wavelengths. The natural greenhouse effect traps heat that makes life on Earth possible. But the *enhanced* effect is warming both the atmosphere and the ocean and changing the dynamics of the water cycle. Climate change poses an immediate and growing threat to California's economy and environment that calls for an urgent response.

Global climate changes are evidenced in ice cores, tree rings, spring snow covers (which have decreased by 10 percent in the Northern Hemisphere), and the melting and collapse of polar ice. The sea level rose about 10 times faster during the twentieth century than during the last 3,000 years. Mountain glaciers are retreating; some of the Sierra Nevada glaciers have lost 70 percent of their volume. Though California's glaciers are relatively small, glacial and ice field melting around the globe contributes to rising sea levels. El Niño events are occurring more frequently. Some plant and animal species have shifted their ranges northward and toward higher elevations, following the temperature and water conditions they need. For the 2014 calendar year, California's average temperature was more than four degrees above the twentieth century average (fig. 74). Across the globe, the year 2014 surpassed 2010 as the warmest year since temperatures began to be recorded in 1880. Parts of the Central Valley and southern state have warmed the most. Summertime heat waves have increased across California, particularly along the North Coast and Mojave Desert regions.

Climate change is altering the planet's water cycle as energy circulates between the atmosphere and the ocean. Water planners rely upon sophisticated climate models to forecast California's changing hydrologic future. The state may experience average annual temperature rises of three to four degrees Fahrenheit in the coming century, with winters five to six degrees warmer and summers up one to two degrees. Higher temperatures will bring more winter precipitation falling as rain instead of snow. The Sierra Nevada snowpack, a vital reservoir, could be

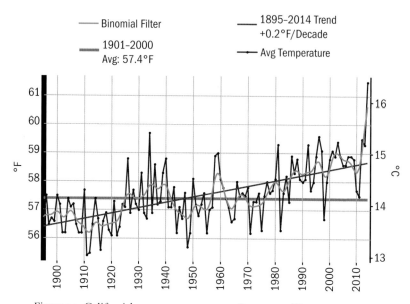

Figure 74. California's average temperature, 1895 to 2014. The year 2014 was the warmest on record, by far (National Oceanic and Atmospheric Administration).

40 percent smaller by the year 2050. Some areas in northern California may actually receive more annual rainfall and bigger storm events, but models suggest the state, as a whole, will be 15 to 35 percent drier by the year 2100. Stream runoff may shift into fall and winter as snow melts, while decreasing during the traditional April to July runoff period. Reservoirs behind dams that have been traditionally operated to handle heavy spring runoff may not have the capacity for rapid storm surges, increasing flood risks. Hydropower generally provides 15 to 20 percent of the state's electricity. Storage challenges will mean less reliable power generation by reservoir power plants. Extreme heat events aggravate crop water stress, increasing irrigation needs and lowering farm production. They also cause electricity

demand to spike in an unfortunate feedback loop aggravating the energy consumption that drives global warming.

Other environmental impacts include less cold water river runoff properly timed for spawning salmon. Large swaths of forest trees have already died due to bark beetle infestations that appear to be tied to warmer temperatures. Fire behavior has become more extreme and the size of the most uncontrollable wildfires has grown since 2000. Though not yet documented, some residents of the Central Valley have noted fewer days of "Tule fog," which matters to fruit growers in the valley because their trees will not produce fruit properly without the colder temperatures fostered by the winter fog.

Because the ocean absorbs about one-quarter of the CO_2 humans put into the atmosphere every year, seawater has become more acidic. Acidic water interferes with shell formation by animals like mollusks (oysters and clams, for example), corals, and plankton, the critically important organisms at the base of ocean food chains. Warming ocean water causes it to expand, explaining much of the sea-level rise that has been observed. In the twentieth century, the sea rose by seven inches, on average, along the California coast. Climate models suggest another 16-inch sea-level rise could happen by 2050, and a rise of 55 inches by 2100, which would bring seawater far inside the Delta, unless river flows increased to push the mixing zone between fresh and saltwater back toward San Francisco Bay. Delta levees will be battered by the rising ocean, threatening islands that are, today, below sea level. Coastal communities and infrastructure at sea level will have to be protected by levees, elevated on stilts, or abandoned.

Assuming humanity decides to promptly take meaningful action against climate change, it will still take a century for all of the long-lived greenhouse gases in the atmosphere to break down. Though the shift to fossil fuel alternatives has not yet happened across the globe, California has a target date of 2020 for lowering the state's greenhouse gas emissions to 1990 levels. Cars and trucks are responsible for over a third of the state's emissions, while electrical power generation and industry account for most of the rest. There has been little progress: since the year 2000, despite a 10 percent increase in population and a 49 percent increase in economic output, greenhouse gas emissions declined slightly (by about two percent). California became the first state with a cap-and-trade system in 2012. Releases of CO_2 by industries were capped (with the 2020 target date in mind) and a finite number of permits issued for those emissions. The permits can be sold by a business that has cut back its emissions, and the purchasing company, then, can legally continue emitting. In January 2015, the governor announced further attacks on climate change by reducing use of petroleum products by half and increasing the use of renewable electricity sources.

Water is a key to this effort. Twelve percent of all California energy consumption goes toward pumping water from the ground; moving water over mountains and through pipes; cleaning it to make it drinkable; or heating and chilling water. The majority of that energy is consumed as electricity and natural gas, and a much smaller amount as oil. "End uses" like residential hot water heaters constitute about 10 percent of California's total energy use, while extraction, conveyance, and treatment make up the other two percent of the state's water-related energy use. Besides the water supply and environmental benefits of conservation, recycling, and reuse, water efficiency reduces the state's energy use and carbon footprint.

EXTINCTION IS FOREVER

Ninety-five percent of California's original wetlands and 89 percent of its riparian woodlands are gone. About 1,400 dams in California convert flowing rivers and streams into reservoirs; in the Sierra Nevada, 600 river-miles have been flooded (map 21). It is no coincidence that California is also the state with the most endangered and threatened species.

Water development and changes in the waterscape are responsible for California's distinction as one of the globe's extinction epicenters of the twentieth century. Dams store water supply, generate electricity, and help control flooding. They also destroy many of the natural processes of rivers and river ecosystems, not just where a reservoir pools water, but also downstream. Dams trap sediments, depriving beaches of sand downstream, along rivers and on the coast. Without sediment deposits and periodic high flows, riparian vegetation is lost, spawning gravels are not maintained, and habitat diminishes for everything in the food chain supported by river ecosystems.

Dams kill fish inside spinning turbines or when fish screens are ineffective at points of diversion. They also change water temperatures downstream. If dams release surface waters that are much warmer than natural flows, fish may become more susceptible to diseases. Some dams send colder-than-normal, oxygen-depleted water downstream from the bottoms of reservoirs. The native fish of the Colorado River adapted to spring floods that provided warm, nutrient-rich waters for spawning. Releases from Colorado River dams, however, are now clear and cold. Inherent in flood control by dams is modification of natural river fluctuations; peak flows are reduced, and that water is

Map 21. California dams (redrawn from California Dams Database, n.d.).

stored for release during dry seasons. Flooding's beneficial effects—scouring sediments, replenishing spawning gravels, and fertilizing floodplains—are lost. Another primary function of most dams is to enable water diversions. Despite state laws that require that enough water remain in streams when diversions occur, to protect fisheries and public trust values, some rivers have been completely dewatered, and most have too little water.

Freshwater entering the ocean is sometimes characterized as a waste; some politicians have declared that no water should be "lost" into the sea. Salmon and other anadromous fish would undoubtedly disagree. They make return trips to inland spawning grounds by tracking the chemical signatures of the river waters of their birthplace as those waters blend into the ocean. Nearly every river or stream that reached the coast of California once supported runs of salmon, Steelhead Trout *(Oncorhynchus mykiss)*, or Pacific Lamprey *(Entosphenus tridentatus)*: fish born in freshwater that migrate to the ocean and return to spawn. California's four distinct runs of Chinook Salmon *(O. tshawytscha)*, also called King Salmon, are named according to the timing of their passage through the Delta as they return from the sea. Winter-run Chinook come back to spawn after one to three years in the ocean. Spring-run adults are most often three years old; unlike those in the other runs, they do not mate and die immediately, but spend the summer high in the watershed, where the water stays cool, and then spawn in late August and September. Fall-run Chinook come in two waves, early and late fall, after spending one to three years in the ocean (fig. 75).

Coho Salmon *(O. kisutch)*, also called Silver Salmon, were once abundant in 582 coastal streams from Monterey Bay up to the Oregon border. Steelhead once used almost every stream along the coast of California. (Steelhead are Rainbow Trout that

Figure 75. Salmon leaping.

have gone to sea—genetically identical, it appears, to the rainbows that live their lives in freshwater streams.)

Californians may wonder why so much media attention has been devoted to these fish in recent years. Why is water for salmon considered particularly important? Or water for another anadromous fish, the tiny Delta Smelt *(Hypomesus transpacificus)?* That two-to-three-inch-long fish was once one of the most common in the Sacramento–San Joaquin Delta estuary. Smelt spend their adult lives in the interface zone where salt and freshwater mix, and move upstream into freshwater to spawn. They are particularly dependent on flows entering the estuary, flows that have been greatly reduced by diversions above and within the Delta. The small fish are also very susceptible to "entrainment," being sucked into diversion pumps. Delta Smelt were listed as threatened under the Endangered Species Act in 1993 and record-low numbers led to "endangered" status in 2010, under the California Endangered Species Act. When large concentrations of smelt come near, pumps have to be shut down. Stopping the pumps reduces supply reliability and water quality for water providers who depend on Delta diversions.

The salmon and steelhead decline has been more noticeably dramatic. Millions once migrated up California's rivers and streams. They were "keystone species" in the rivers and their loss became one measure of the greater loss of wildlife and riparian habitat in California. In the winter of 1883 to 1884, over 700,000 salmon were caught for commercial sale along the Delta. In that era, voracious canneries depleted the vast salmon populations, but it was the dams constructed during the 1940s, 1950s, and 1960s that dealt the near-fatal blows to the species. The Sierra Nevada Ecosystem Project, a 1996 study for Congress, found that only about 676 miles of stream habitat for salmon remained accessible, of 1,774 miles once used in the Central Valley and Sierra Nevada. More critical than lost stream length was the inability to reach suitable spawning sites in the upper reaches of those streams; only 10 percent of that habitat remains (map 22).

In the Sacramento River system, just 191 winter-run Chinook returned in 1991. Spring-run numbers were down to 2,300 in 1992. The winter-run has been listed as endangered and the spring-run as threatened. Fall-run Chinook are the only salmon that are now present in both the Sacramento and the San Joaquin watersheds; this run is also threatened. Coho Salmon, estimated at one million in the mid-1800s, numbered less than 6,000 in coastal streams by 1996. The population south of San Francisco Bay is endangered. Other genetically distinct Coho populations in Northern and Central California are threatened. Steelhead Trout are classed as endangered along the South Coast. Genetically distinct populations on the Central and North Coasts are threatened.

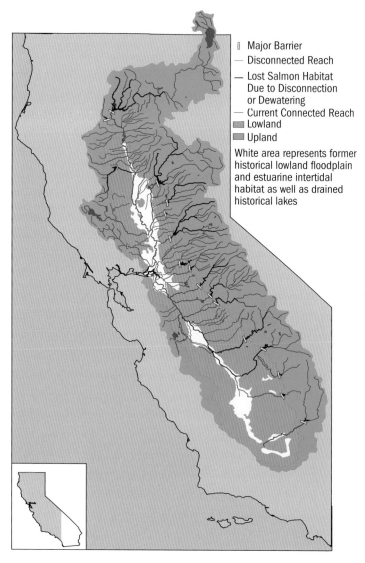

Legend:

- Ⅱ Major Barrier
- — Disconnected Reach
- — Lost Salmon Habitat Due to Disconnection or Dewatering
- — Current Connected Reach
- Lowland
- Upland

White area represents former historical lowland floodplain and estuarine intertidal habitat as well as drained historical lakes

Map 22. Disconnected salmon rivers (redrawn from Bay Institute of San Francisco 1998).

Salmon are amazing water creatures. Born in our rivers, they mature in the ocean and return to their birthplaces to reproduce. During their time offshore, they may travel more than 2,000 miles across the ocean; some California salmon reach the coast of Siberia. As each returns toward its home estuary, it somehow detects the "fingerprint" scent of *the* river that it must follow back to its birthplace. Salmon noses can detect the identifying substances in concentrations as low as one part per 3,000,000,000,000,000,000,000 (3×10^{18})!

After entering the river, salmon stop feeding. Their entire metabolism develops "tunnel vision," focused on traveling and the ultimate reproductive goal. Their digestive tracts break down. Stored fat and muscle are converted into energy to power their upstream battle against the current. They run a gantlet of predators, including bears, osprey, and, for thousands of years, Native American fishermen. They leap over cascades and physical obstacles. If they successfully pass a dam by navigating a fish ladder, they may still become confused by the slack water of a reservoir, because the push upriver, against a current, is part of their genetic hardwiring.

A female that reaches her home destination looks for clean gravel with a flow of oxygenated water passing through. She digs a nest and releases hundreds of eggs while a nearby male fertilizes them. Somehow, as gametes fuse, the new life records the chemical signature of the home water. When parent fish are intercepted during migration and taken to a hatchery on another creek, their offspring, raised in chemically different water, return to the hatchery creek rather than follow the traditional route.

Parents die soon after spawning. Their large decomposing carcasses are gifts of food for their offspring and for many insects, other fish, mammals, and birds. The great biomass of salmon flesh is essential to many species along the river. Fry that emerge from eggs may spend more than a year in the home stream or head immediately downstream for the sea, depending on the species. They need unpolluted water that stays cool in the shade of riparian vegetation. Many of the insects that salmon eat in the river also depend on willows, cottonwoods, and similar shoreline plants to complete their life cycles. Young smolt pass, in reverse, the same gantlet of predators and barriers that their parents ran. As they finally reach salt water, they undergo physical adaptations to the ocean.

As dams and diversion canals were planned and constructed, the obstacles they presented to fish *were* anticipated. Fish screens were designed to fend fish off from the diversion intakes to canals. Fish ladders were constructed, and hatcheries were built (fig. 76). The California Department of Fish and Wildlife operates eight salmon and Steelhead hatcheries, and the U.S. Fish and Wildlife Service operates two hatcheries in California (map 23). Although these efforts are well intentioned and of some benefit, the declining numbers of fish suggest the ineffectiveness of many of the older fish screens, ladders, and hatcheries. What migrating fish need most is an adequate share of free-flowing water.

That lesson was suggested at Butte Creek, one of the Sacramento River tributaries. In 1987, a drought year, only 14 Chinook Salmon returned to spawn there. In 1988, the year after small dams along Butte Creek were removed, 20,000 Chinook made it to the spawning grounds.

Conflicts between water for fish and water for farms generated a public outcry during the summer of 2001 on the border between Oregon and California. The U.S. Bureau of Reclamation, faced with a drought and needing to maintain minimum flows in the Klamath River to protect endangered species of fish, told farmers

Figure 76. Salmon being sorted for artificial spawning at the Feather River Fish Hatchery.

Map 23. California fish hatcheries (redrawn from California Department of Fish and Wildlife, n.d.).

they had to cut irrigation water by 90 percent. "Dreams Dry Up in Klamath Basin," the *Los Angeles Times* headline read on July 23, 2001. The story called "the unfolding tragedy...the culmination of a century of unsustainable federal policies designed to satisfy demands for cropland, fishing, population growth and wildlife protection." The human issue was more complex than just "farmers versus fish." A major fishing and canning industry for *other* humans, on the coast, depended on the Klamath River salmon runs. The fish had also been the cultural lifeblood and dietary staple of Hoopa and Yurok Indians living along the river. The Klamath Hydroelectric Settlement Agreement of 2010 aimed at restoring salmon populations to 25 percent of historic levels (250,000 adult spawning salmon and steelhead) by removing four dams and opening up 300 miles of spawning and rearing habitat (see the "Raze Existing Dams?" section, Chapter 5).

Over-allocation of the Klamath River system had parallels to the problems of the San Joaquin and Colorado River systems, where consequences of unrealistic water allocations and unsustainable population growth were also being realized. The reason Californians were "suddenly" hearing so much about salmon and smelt, water conflicts, and endangered species was that resources had been stretched beyond natural limits (fig. 77).

The Endangered Species Act is societal recognition that human actions should not cause any species to disappear forever from the Earth. Enforcement of habitat protections that follow listing under the Act can be controversial and feel onerous to people whose activities are affected. But enforcing the Act requires ending the threats. Other laws also protect water-dependent species, but lack of effective enforcement (often due to underfunding) has been a problem in California's water management.

Figure 77. A close-up look at salmon at the Nimbus Fish
Hatchery during the American River Salmon Festival.

Four of the tributary streams feeding Mono Lake were completely dried by diversion into the Los Angeles Aqueduct. Section 5937 of the state Fish and Game Code, adopted in 1937, reads,
in part, "The owner of any dam shall allow sufficient water at all
times to pass through a fishway, or in the absence of a fishway,
allow sufficient water to pass over, around, or through the dam, to
keep in good condition any fish that may be planted or exist
below the dam." This mandatory law uses the word "shall," rather
than "may." Court tests have affirmed that the law applies to all
dams, regardless of when they were approved or constructed.

Correcting violations of this law was part of the environmental victory at Mono Lake that, in 1994, produced a protective

plan approved by the State Water Resources Control Board (SWRCB). In 1995, the SWRCB also affirmed, in *California Trout, Inc. v. Big Bear Municipal Water District*, that "the protection of fisheries from inadequate flows is necessary regardless of where fish live, be it in the largest navigable river or the smallest mountain stream." That ruling stated, "Fish were granted that protective right the instant California became a state in the Union."

Although a half million spring-run Chinook migrated up the San Joaquin River each year before Friant Dam was built by the Central Valley Project (CVP), about 60 miles of the San Joaquin were dewatered after 1942. Correcting that violation of state law has begun. Meanwhile, portions of every stream in the Tulare Basin remain dry, the water running elsewhere through pipes and irrigation canals. The express purpose of Pine Flat Dam, on the Kings River, was to keep water from ever reaching its traditional terminus in Tulare Lake, so crops could be grown on the dry lakebed.

A THIRSTY GARDEN

The food each of us eats each day represents an investment of 4,500 gallons of water, according to the California Farm Bureau. California is the number one agricultural state in the nation, with nine of the top 10 agricultural counties in the United States. The state's agricultural production was valued at $45 billion in 2012. Fifty-five percent of the nation's produce is grown here. California is the nearly exclusive source of special crops such as almonds, artichokes, dates, figs, olives, and raisins. About 73 percent of the state's agricultural revenues are from crops. If urban communities are the "backbone" of the California economy, farms might be considered its vital organs. With 80 percent of the state's developed water going to agriculture (ranging between 35 MAF to 45 MAF a year), and with urban interests ever thirsty, how efficiently farmers use their portion is a major concern.

Although agricultural land is steadily being lost to urban encroachment, in 2009 the state still had 43 million acres of agricultural land, about 40 percent of its land area. Of this, 31 million acres are grazing land (more than half of that on public lands) and 12 million acres are cropland (nine million of those acres are irrigated). There are almost 80,000 farms and ranches, but the great majority of farmland is divided among about 5,000 particularly large landholdings.

In California, land and water rights held by farmers have historically made agricultural property attractive to urban developers. Los Angeles County was the leading agricultural county in the nation until the 1950s. The Santa Clara Valley, today's fully urbanized "Silicon Valley," was also once a major producer of fruits and vegetables. Repeating the pattern, the Central Valley is now under heavy pressure from urban sprawl. Every year, 15,000 acres of agricultural land in the state succumb to concrete and asphalt. Farms are also being pulled out of production when urban suppliers purchase agricultural water.

Water availability controls crop choices and irrigation methods. Much of the water used by agriculture in the state goes to five crops: alfalfa, nuts, rice, grapes, and irrigated pasture. Grapes grown with drip irrigation may use one to two feet of water on each acre (fig. 78). Cotton, grown in the same location with flooded furrows, would require three to four acre-feet. Pressurized drip and microspray systems are efficient and work well for some crops but are not as effective on others (figs. 79, 80).

Over-irrigation can put runoff water, carrying salts, fertilizers, and pesticides, into groundwater basins and surface water. If

Figure 78. Drip irrigation of grapes.

Figure 79. Orchard irrigation with groundwater pumping.

Figure 80. Vegetable crops flood-irrigated with siphon tubes from a canal.

fertilizer, which augments the nitrogen essential for plant growth, is applied more heavily than can be used by plants, the excess can leach into groundwater. Nitrates in drinking water pose an especially acute risk to infants, causing birth defects and, with long-term exposure, risks of gastric, esophageal, or ovarian cancers or non-Hodgkin's lymphoma. These health impacts have been disproportionately felt by the poorer Latino communities of the Central Valley.

Agricultural runoff comes not solely from farm fields growing plant crops. California rangelands are drained by 9,000 miles of waterways. Cattle love to hang out in the shade offered by trees that line streams and ponds, and they also like to stay close to drinking water. Their urine and feces accumulate there and introduce nitrates, ammonia, and fecal bacteria that must be kept out of water supplies. In stream and pond environments, nitrates promote algae growth. Ammonia, from urine, can be toxic to many aquatic creatures. Concentrations of cattle near water have physical consequences for water quality. Where riparian vegetation is eaten away or trampled, shade is reduced and water temperatures can become too high for fish. Streambanks may also become unstable. Destabilized streams can form deep cuts that drain groundwater out of nearby meadows. Overgrazing of a watershed can increase runoff and lead to flooding that carries sediments into water sources. Ranchers can minimize these effects by fencing cattle out of streams and riparian areas and alternating short-duration, intense grazing with rest periods.

Dairy products provide the highest revenue on the list of California's farm commodities. Dairies and cattle feedlots also generate enormous concentrations of waste. Contrary to the cute images in television ads that promote California's dairy products, most modern dairies do not allow their cows to wander

Figure 81. Up to her knees in muck; the polluted water must be properly managed by dairy farmers.

idyllic pastures (fig. 81). The Chino Basin, on the upper watershed of the Santa Ana River, in the "Inland Empire" counties of Riverside and San Bernardino, once had the highest concentration of dairies anywhere in the world, with 300 facilities holding up to 40 cows per acre. In 1995, 329 dairies operated there with over 310,000 dairy cows. By 2012, the number of operating dairies dropped to 111 with 115,000 dairy cows. Milk production had fled conflicts that came with rapid urbanization, so that, in 2014, almost three-quarter of the state's dairies were in the Central Valley counties of Tulare, Merced, Kings, Stanislaus, and Kern.

The SWRCB and the nine regional boards require dairies to isolate watering and feeding areas from streams and to divert rainfall runoff away from manure areas. They also require dairies to recycle manure, urine, and the water used to hose down facilities, storing it in leak-proof lagoons until it is used to irrigate and fertilize pastures. Lagoons have been known to leak or overflow, however, when not well designed and maintained, and they release noxious gases. Applying manure from the lagoons

Figure 82. A waste "lagoon" for gathering fertilizer and urine runoff from a dairy in the Santa Ana River watershed.

Figure 83. Flooded California rice fields.

to fields has to be done no faster than crops can absorb it, to avoid polluting streams or groundwater (fig. 82).

California is a major rice-growing state. Though rice fields must be flooded, rice actually takes less water to grow *per serving* than almost any other crop in the state. Rice fields also substitute for some of the lost wetlands habitat in the Central Valley, benefiting migratory waterfowl (fig. 83).

That kind of environmental mitigation was not always the norm. The standard "reclamation" practices of the nineteenth and early twentieth centuries aimed primarily to convert "worthless" mosquito-infested wetlands into productive farmland. Agricultural expansion was behind many habitat losses in California. Farms also became a major contributor to water pollution in the state's surface and groundwater ecosystems. Yet, for 20 years, farmers in California were exempted from Clean Water Act (CWA) controls on insecticides, herbicides, and other pollutants that entered the state's waters. One dormant-spray pesticide, Diazinon, used in Central Valley almond orchards, began reaching San Francisco Bay during high-runoff episodes in concentrations over 100 times the allowable limits. The waiver ended in 2002, as the SWRCB issued the first statewide "general permit" controlling irrigation return waters and stormwater runoff from farms.

Crops separate irrigation water from most of the salts it carries, leaving the minerals concentrated in the soil. To keep salination from increasing to the point where no plant can survive, farmers must apply more water than the plants need. The extra water flushes salts down into the groundwater below. In places where topsoil overlies impermeable layers, excess water and salts cannot drain away. When they accumulate and reach back into the root zone, farming may become impossible. Around the world, including in the Fertile Crescent of the Middle East, salination has been the historic bane of irrigated agriculture.

Different crops have different tolerances for soil salinity. Strawberries are very sensitive. Cotton, alfalfa, pomegranates, and pistachios are salt-tolerant commercial crops. Some plants, such as eucalyptus trees, actually "mine" salts. The Westlands Water District, on the west side of the southern San Joaquin

Valley, planted eucalyptus trees at the edges of fields, hoping they would absorb salty drainage water, but with limited success. The 600,000-acre district has some of the greatest salinity problems in California. That side of the valley has soils created by marine sediments that were naturally saline. An extensive clay layer below the topsoil also blocks deep percolation of irrigation water. On the east side of the valley, by contrast, granitic soils drain well and are naturally low in mineral salts.

Water sources used for irrigation also vary in their salinity. The Sacramento River carries about 270 pounds of salts in each AF, compared to 2,000 pounds in the Colorado River. The diminished flows of the lower San Joaquin River carry 1,200 pounds per acre-foot (and up to 3,300 pounds at times). A total of 1.3 million tons of salt annually enters the Delta from the San Joaquin River.

To drain off saline waters, farmers bury perforated pipes, called tiling, at spaced intervals beneath fields. Sometimes the runoff gathered this way goes into flowing rivers and the ocean. In other places, it goes to local evaporation ponds (fig. 84).

The San Joaquin Valley's drainage problems were anticipated when the CVP was designed. The San Luis Drain was planned to carry water northward through the valley all the way to Suisun Bay. Eighty-five miles of the drain were built, beginning in the south. By 1973, construction reached as far as Kesterson National Wildlife Refuge, near Gustine. Concerns about the impacts of sending the polluted water into the Delta halted construction, forcing the water to remain at Kesterson. In 1983, after a decade of drainage water ponding there, thousands of waterfowl and shorebirds began to be born with deformities and to die. Bird embryos had protruding brains, missing eyes, twisted bills, and grossly deformed legs and wings (figs. 85a, 85b). Elevated selenium levels were responsible.

Figure 84. A drainage canal for returned agricultural water in the San Joaquin Valley.

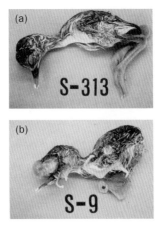

Figure 85. Black-neck Stilt *(Himantopus mexicanus)* embryos, (a) normal (S-313) and (b) deformed (S-9), from a Kesterson Reservoir nest, in 1985.

Selenium is a trace element found in much of the Westland Water District's soil. It is necessary for life but, in a process called biomagnification, can become concentrated to toxic levels as it ascends a food chain. Not just birds were affected; toxic levels of the substance showed up in insects, frogs, snakes, and mammals (concentrated selenium can also be toxic to humans). The drain to Kesterson was closed in 1986. Selenium in farm runoff and evaporation ponds across the state has been closely controlled since that time. A drainage solution had been promised to farmers receiving CVP water, however, and court decisions confirmed that some solution must be provided. The Delta outflow option was not viable because salts would impact water headed for so many domestic users and the fragile estuary ecosystem. Another option, piping drainage water directly westward to the ocean, could have caused major environmental impacts in the Monterey Bay National Marine Sanctuary.

Nearly three decades after the Kesterson drain was closed, a 2015 settlement has been drafted between Westlands and the Bureau of Reclamation. Congressional legislation will have to follow for the controversial plan to take effect. If the deal goes through, Westlands would take over responsibility for the selenium and salt cleanup, relieving the federal government of its responsibility for a drainage solution, saving taxpayers an estimated $2.7 billion. In return, Westlands would be forgiven $342 million the water district still owed for CVP construction. Only 100,000 of the district's farm acres prone to selenium runoff would be fallowed, although a Bureau of Reclamation study had called for three times that many to be taken from production. A controversial element in the deal would grant the district a permanent water contract, no longer requiring renewals at 25-year intervals, with a higher water delivery priority. Westland's contracts for water have been "junior" to those with riparian and pre-1914 appropriation rights, so in drought periods they have been last in line for CVP water. With this deal, their contract allocation of CVP water would be capped at 889,800 AF.

The big question is whether Westlands will actually solve the drainage issue, considering that the federal government has not done so. Some Westlands farmers that recently switched from flood irrigating vegetable crops to drip irrigation on grapevines and nut trees are applying much less water, thus reducing the amount of contaminated drain water. A model farm system has been developed in the district, progressively reusing its irrigation water on a succession of salt-tolerant crops, including forage grasses and cactus. A solar powered desalination plant has been tested in the neighboring Panoche Water District to process farmland drainage and produce distilled water, solid salts, and minerals with industrial value. A larger-scale plant, though only producing 2,240 AF per year, is scheduled for construction in 2015. Though it is intriguing to see inland desalination plants powered by the sun, instead of by fossil fuels, the Westlands Water District's thirst requires almost 900,000 AF per year.

That thirst, and the competition for limited water out of the Sacramento–San Joaquin Delta, became rigidly demanding as San Joaquin Valley growers converted from vegetable crops and cotton to higher value nut orchards. Unlike annual crops, where a growing season might be skipped now and then during droughts, long-lived trees must be kept alive with irrigation water. Almond trees are particularly thirsty. In 1995, California produced 370 million pounds of almonds, but by 2013, the state's farms were harvesting *2 billion* pounds. Each individual nut represents one gallon of irrigation water. Acreage devoted to almond orchards expanded by 20 percent between 2007 and 2013, and totaled 940,000 acres in 2015. Similarly, pistachio acreage has grown by 75 percent. California accounts for 80 percent of the almonds and about 40 percent of the pistachio nuts consumed around the world.

Even without the contributions of the San Luis Drain, San Joaquin River flows that approach the Delta today consist of little but agricultural drainage water. The Regional Water Quality Control Board enforces Total Maximum Daily Load (TMDL) requirements for contaminants in the state's waters. A TMDL specifies the maximum amount of a pollutant that a water body can receive before being declared impaired or polluted, and allocates pollutant "loadings" among point and nonpoint sources. Tougher discharge standards for the San Joaquin must be complied with by those CVP exchange contractors whose water supply out of the Delta is saltier than the water that originally flowed to them out of the mountains, down the San Joaquin River.

With discharge standards tightening, "Grassland Area" farms, on 97,000 acres on the west side of the San Joaquin Valley near Los Banos, initiated a regional drainage effort. Their goal became finding ways to keep from delivering any salty water to the San Joaquin River. With tiling in place beneath about 30,000 acres, by 2002, subsurface water was collected and used to irrigate salt-tolerant crops. Asparagus, alfalfa, pistachios, pomegranates, and Bermuda grass were grown. Salts, inevitably, will continue to concentrate. After several decades, the farmers will have to deal with permanent storage of very concentrated slurries.

Farther south, on the former bed of Tulare Lake, from which no rivers drain, 44,000 farm acres have constant salinity issues. Tilewater is gathered and delivered to 4,700 acres of evaporation ponds. Unfortunately, as at Kesterson, such ponds look quite attractive to birds. The law now requires that evaporation ponds be designed to discourage use by birds (with steep walls and frightening sounds and calls). Nearby alternative wetlands and nesting habitats, with features meant to *attract* birds, are maintained with clean water.

ASKING TOO MUCH OF THE COLORADO
RIVER AND THE SALTON SEA

In February 2008, researchers with the Scripps Institute analyzed the water budgets of the Colorado River's two largest reservoirs, Lake Mead and Lake Powell. Human demands for water from the reservoir system, along with predicted runoff declines and evaporation increases due to global warming, would, they predicted, produce a 50 percent chance that functional storage levels in the two reservoirs would be gone by 2021. There was a 50 percent chance that the minimum levels for hydroelectric power generation would be reached in both lakes in 2017. Some methods used in this study were criticized, but the findings were startling enough to draw national media attention and raise awareness that there are limits to this over-allocated river system.

After 14 years of drought exacerbated the problems, a new Colorado River System Conservation Program was announced in 2014 by Metropolitan Water District of Southern California (MWD), Las Vegas, Phoenix, Denver, and the Bureau of Reclamation. They created a fund that will pay farmers, industries and cities to conserve water across the watershed, aiming to keep the Mead and Powell reservoirs high enough to avoid a formal water shortage declaration. A shortage will be declared if Lake Mead's elevation drops below 1,075 feet above sea level. Early in 2015, Mead's elevation had fallen to 1,089 feet. If the new program is not effective at curbing water use, Lake Mead's elevation could drop below 1,000 feet in the next decade, reducing hydroelectric capacity, falling below the intake that serves Las Vegas, and severely impacting all Colorado River users.

Details for transferring Colorado River water rights from Imperial Valley farms to San Diego County were finalized back in October 2003, in a Quantification Settlement Agreement to comply with the requirement for California to stop taking more than its allotted water right from the river. The agreement required the state to take responsibility for impacts on the Salton Sea as less water drained off farms. In a January 2008 environmental impact report, the Department of Water Resources described a preferred alternative that would leave a smaller hypersaline portion, along with a larger section of the sea maintained with salinity low enough to support fish (and fish-eating birds). Although California voters approved a bond in 2006 authorizing $46 million toward Salton Sea restoration and Congress authorized $30 million in matching federal funds, the state's budget deficit stalled that effort, which remained unresolved in 2015.

Imperial Valley drainage issues present a particular dilemma in the southern desert. In the valley, 38,000 miles of subsurface drainage pipes gather water from 500,000 irrigated acres. The Salton Sea is the "evaporation pond" for that water, taking in more than four million tons of dissolved salt and tens of thousands of tons of fertilizers each year. It is now 25 percent saltier than the ocean. Yet 400 million fish still reside in its harsh water, making it the most productive fishery in the United States.

As wetland habitat disappeared from southern California's coast, the Salton Sea became an essential alternative site for millions of migratory birds—a critical rest and feeding spot along the Pacific Flyway. Bird tagging has revealed that Salton Sea birds fly everywhere across North America. Almost the entire population of Eared Grebes *(Podiceps nigricollis)* in North America uses the sea, and it serves endangered species such as the Brown Pelican *(Pelecanus occidentalis)*, Yuma Clapper Rail *(Rallus longirostris)*, Black Rail *(Laterallus jamaicensis)*, and Greater Sandhill Crane *(Grus canadensis)*. Birds take advantage of the sea's enormous

Figure 86. The Salton sea, teeming with life but a place of disturbing mortality.

productivity. The food chain begins with algae fertilized by farm runoff nutrients. Hot weather algae blooms, however, lead to algae mortality and decomposition that can consume more oxygen than photosynthesis generates. Oxygen-depleted water then causes fish to suffocate. Fish die-offs may be responsible for an assortment of diseases and death in bird populations. Though it still teems with life, paradoxically, the changing sea is also a place of widespread, disturbing mortality (figs. 86, 87).

Figure 87. The bones from tilapia die-offs covering the beach at the Salton Sea.

Furthermore, salinity levels are approaching the limits for any fish. If no freshwater entered to replace evaporation, the sea could lose its fish in another decade. It takes about 1.5 MAF to replace evaporation off the sea each year. Should less water enter, the sea would decline and rapidly concentrate the salts it already contains. To stabilize the salinity level would require not only enough farm runoff to replace evaporated water, but also removal of four million tons of salt sent to the sea from Imperial Valley farms each year.

Exacerbating efforts to address the Salton Sea's progressive salination, coastal cities have plans to purchase water from Imperial Valley farmers to service more population growth and development. Although the farm drainage water gradually increases salinity in the sea, losing river water to those cities would cause a more rapid decline in water level and further salt-water concentration.

The Salton Sea has sometimes been dismissed as a "man-made mistake." Yet the below-sea-level basin flooded many times in the past, whenever the Colorado River shifted its course that direction. The physical size of the sea means it will not simply stop drawing migratory birds. Allowing it to completely dry up would leave birds with even less habitat in a state that has paved over most of its wetlands. Drying up the sea would also expose a vast source of fine particulate dust, creating air pollution problems far greater than those at Owens Lake. Dust clouds off the lakebed would have multi-county impacts.

Imperial Valley farming, though incredibly productive in a climate that allows multiple harvests each year, is sometimes dismissed as a waste of water in that desert environment. About half the land in Imperial Valley grows alfalfa, a crop that uses twice as much irrigation water there as in other parts of the state. If up to 25 percent of the farmland were to be fallowed, as has been suggested, Colorado River water not used on that acreage *could* be allocated directly to the Salton Sea. That kind of decision, a water right directly tied to the environment, would signal a major shift in public attitude and water management in California.

OUT OF SIGHT, OUT OF CONTROL

Although groundwater interacts with surface water, and California has laws to govern every aspect of surface water use and quality, until 2014 there were no statewide groundwater management laws. Land ownership, in most cases, brought with it the right to essentially unregulated pumping of groundwater, until 2014, when the Sustainable Groundwater Management Act was signed into law (see the "Sustainable Groundwater" section, Chapter 5). The California Constitution stipulates that water is owned by the people and must not be wasted or put to an unreasonable use by those who have water rights. Beyond that overall requirement, groundwater, unlike surface water, had not been under statewide control after 1850. Historian Norris Hundley Jr. characterized California's first century-and-a-half of groundwater "management" as a "chaotic and environmentally destructive practice of management by numerous local water districts and agencies – a practice that has meant no management at all" (2001, 530) (map 24).

A clear example occurred as urban sprawl spread out of Southern California's coastal basin into the Mojave Desert. The Mojave River groundwater basin produces a net annual supply of 50,900 AF, but the pumping rights of basin landowners totaled 280,000 AF. Priority rights go to those who acquired land first.

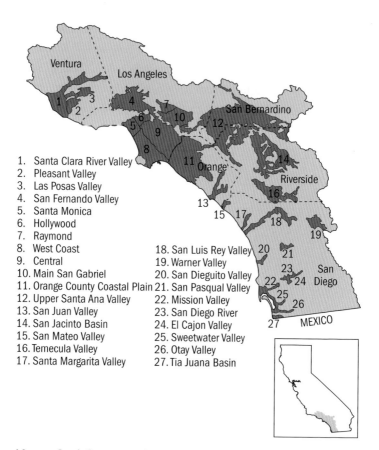

1. Santa Clara River Valley
2. Pleasant Valley
3. Las Posas Valley
4. San Fernando Valley
5. Santa Monica
6. Hollywood
7. Raymond
8. West Coast
9. Central
10. Main San Gabriel
11. Orange County Coastal Plain
12. Upper Santa Ana Valley
13. San Juan Valley
14. San Jacinto Basin
15. San Mateo Valley
16. Temecula Valley
17. Santa Margarita Valley
18. San Luis Rey Valley
19. Warner Valley
20. San Dieguito Valley
21. San Pasqual Valley
22. Mission Valley
23. San Diego River
24. El Cajon Valley
25. Sweetwater Valley
26. Otay Valley
27. Tia Juana Basin

Map 24. South Coast groundwater basins (redrawn from California Department of Water Resources 1998).

Meanwhile, every new resident in the sprawling developments exacerbated the problem, because with land ownership came the right to pump groundwater. A contentious adjudication process began in 1990, involved a series of court reviews, and was finally settled in 2002.

Statewide, pumps annually remove between 500,000 AF and 1.5 MAF of groundwater a year, with the greatest losses from the

Tulare Lake and San Joaquin basins. NASA has developed remote sensing satellite systems for monitoring both groundwater levels and land subsidence from space. Between 2011 and 2014, groundwater levels fell more than 100 feet in places, as the Sacramento and San Joaquin river basins lost 4 trillion gallons of water every year.

Overdrafting can lead to land subsidence (map 25). As groundwater extraction empties the pores between particles, fine-grained sediments are compacted. Subsidence generally does not mean that the aquifer cannot later be recharged. Though fine-grained sediment compaction is often not reversible, most of the storage capacity of aquifers is not in such fine material, but in coarser sediments. Land subsidence is primarily a threat to roads, buildings, canals, and low-lying coastal areas that may be flooded by seawater. Overpumping in Santa Clara County in the 1930s caused the ground beneath San Jose to sink so far that it was threatened with flooding during high tides in San Francisco Bay. When the South Bay Aqueduct came on line in 1965, the local water district purchased water from the SWP to reduce groundwater dependence.

Subsidence was a great problem in the first half of the twentieth century. Correcting San Joaquin Valley overdrafting became a primary purpose of the CVP (fig. 88). But subsidence became noticeable again during the drought of 1987 to 1993 and in every drought cycle since as farmers relied more heavily on wells. At Edwards Air Force Base, where the space shuttle lands in the desert north of Los Angeles, subsidence has caused 12-foot-deep, 2,000-foot-long fissures on a runway. Deformation of the surface has resulted in the need for repairs to the California Aqueduct. The original siting of the California High Speed Rail was discovered to be directly through a large subsidence bowl.

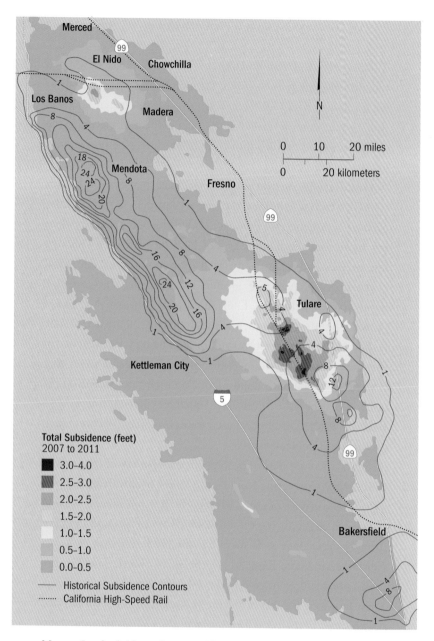

Map 25. Land subsidence from groundwater pumping in the San Joaquin valley, January 2007 to March 2011 (from NASA-JPL radar analysis; redrawn from Borchers and Carpenter 2014).

Figure 88. Near Mendota, in the San Joaquin Valley. The land subsided nearly 30 feet here due to groundwater pumping between 1925 and 1977.

Groundwater pumping can also lead to seawater intrusion. As in the Bay-Delta, where fresh river flows hold back salt water, removing freshwater from the water table may allow seawater to intrude far inland through the ground. In the Salinas Valley, seawater moved six miles inland into a 180-foot-deep aquifer and two miles inland into a 400-foot-deep aquifer. Orange County began percolating water in settling basins and injecting it through wells back in the 1950s, once Colorado River water imports became available. This and other coastal water districts successfully created freshwater barriers to block seawater from contaminating domestic water supply wells.

Bad News beneath Your Feet

We cannot know all of the bad news hidden beneath our feet, the product of our casual disregard or mistaken belief that the ground can clean anything. As we have learned, actual cleansing of contaminated aquifers is costly and difficult. Some contamination is impossible to correct; only time, measured in centuries, may bring a solution. The city of Fresno learned that bitter lesson when 50 of its 300 domestic wells were contaminated by a pesticide and had to be closed.

Industrial and mining contamination reflects water's natural talents as a solvent. Some of the pollution has been with us since the gold rush or before. Mercury was widely used by the '49ers and the miners who followed them to separate gold from ore. The toxin is still present in Gold Country rivers. Hardrock mines came later; today, drainage water emerges from many of them as strong as battery acid (fig. 89). Arsenic, a naturally occurring carcinogen, made national news when the Clinton administration adopted more stringent drinking-water standards that the Bush administration first set aside, then later adopted.

Other contaminants are as modern as rockets and the atomic bomb. Radioactive slag heaps from uranium mines in the Colorado River Basin above Lake Mead threaten that water supply for Southern California. Perchlorate, a rocket fuel that affects thyroid glands, is moving in underground plumes that may take thousands of years to reach discharge areas. The Environmental Protection Agency (EPA) announced, in July 2001, a 240-year "restoration plan" for groundwater contaminated with perchlo-

Figure 89. Acidic mine drainage in the Sacramento River headwaters.

rate from an Aerojet site near Rancho Cordova. The Sacramento suburb has nine wells with undrinkable water and 13 others at risk from the spreading plume. Perchlorate plumes in Arizona also threaten the Colorado River. Perchlorate contamination from a Lockheed plant affects groundwater wells in Rialto, Mentone, Loma Linda, Redlands, and Riverside, all in Riverside County.

There are numerous other examples of industrial groundwater pollutants. The State Department of Health Services identified 37 wells (out of 246 tested) in the San Gabriel Valley that were contaminated with an industrial solvent, trichloroethylene (TCE), also used by Aerojet. Chromium-6 forced the city of Glendale to stop using groundwater for half of its supply and switch solely to surface water sources. The movie *Erin Brockovich*, about a groundwater contamination lawsuit against Pacific

Figure 90. Water-filled excavations marking where leaky gas tanks must be replaced. Decontaminating the groundwater itself is far more difficult.

Gas & Electric over chromium-6 used in a power plant, broadened public awareness of the health issues and consequences of such contamination. The city of Los Angeles has undertaken costly measures to clean up contaminated groundwater basins.

Methyl tertiary butyl ether (MTBE) is a gasoline additive that was intended to reduce air pollution by improving fuel efficiency. It has an incredible affinity for water, easily dissolves, and spreads into groundwater wherever fuel tanks leak (fig. 90). MTBE is a carcinogen that gives water an unpleasant turpentine smell and taste. Until it was banned, boat engines introduced it into reservoirs and lakes. Anyone who has ever looked closely at the water near a marina where boats operate has seen gasoline and oil scum mirroring the surface. Two-stroke carbureted engines of jet skis, or "personal watercraft," were notorious for their inefficiency; such engines dumped a gallon of gasoline directly into the water for every five gallons burned. In 2001, all two-stroke engines (not just jet skis) were banned at Lake Tahoe and at Whiskeytown National Recreation Area. By the end of 2003, all MTBE was phased out and replaced by ethanol as a fuel efficiency additive in gasoline in California.

Fracking

The twenty-first century brought a boom in oil and natural gas extraction across the nation, including in California, as directional drilling techniques combined with high-volume hydraulic fracturing, or "fracking," were developed to stimulate the bedrock to release fossil fuels that traditional pumping could not extract. Fracking uses large volumes of water mixed with chemicals pumped underground at high pressures. A related step is "acidizing," which injects hydrofluoric and other acids to dissolve shale rock. Some of the water returns to the surface and can be reused, but much of it is toxic water that is stashed deep underground via waste wells. One of the most disturbing aspects of fracking is that amount of "lost" water that, when reinjected into the depths, becomes locked away from the active water cycle.

Energy developers became very excited about the potential for oil and gas extraction from California's Monterey Shale formation, which underlies much of the Central Valley's farmland and groundwater aquifers. The formation is also found in Los Angeles and Santa Barbara Counties, where it overlaps with dense human populations (map 26). Originally estimated to hold over 15 billion barrels of oil, about two-thirds of the nation's shale oil reserves, some of the excitement waned after May 2014 when the U.S. Energy Information Administration slashed, by 96 percent, the

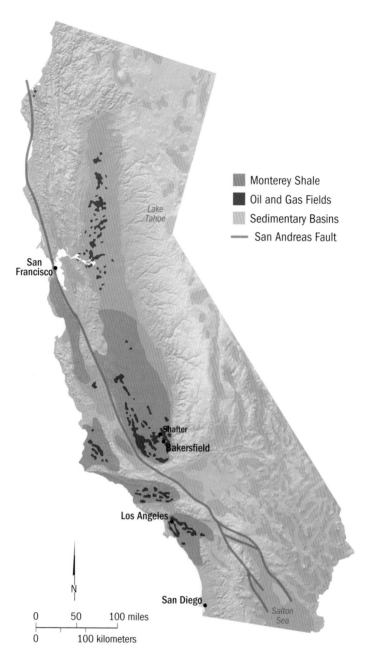

Map 26. Monterey Shale Formation and Oil and Gas Fields in California (redrawn from Seeley 2014).

estimated amount of deposits that will actually be recoverable. Because of complex geological layering, fracking cannot be as effective there as elsewhere in the nation. Still, 600 million barrels of oil will be extractable with existing technology.

Wells that have been fracked in California, as of 2015, were primarily located in Kern (2,361 wells), Los Angeles (124 wells), and Ventura (456) Counties, with smaller numbers in nine other counties, including Monterey, Fresno, and Santa Barbara. About 5.4 million Californians live within a mile of one of the state's 84,000 existing oil and gas wells. A single fracking rig may use 1.46 million gallons of water every week.

In July 2014, the California Division of Oil, Gas, and Geothermal Resources (DOGGR) ordered energy companies to stop injecting fracking wastewater into protected groundwater aquifers near Bakersfield, because it posed a danger to health, property, and natural resources. Seven injection wells were supposed to be depositing toxic wastewater into "sacrifice aquifers"—classified as too deep or too poor in quality for use as drinking or irrigation water were—but were, instead, likely contaminating protected groundwater supplies. This was in the same region, where agricultural pumps had been drilling deeper and deeper during years of extreme drought. Regional water officials learned about the problem while scrutinizing another concern over unlined surface sumps and drilling pits.

That news came several years after a 2011 audit commissioned by the US EPA had faulted DOGGR's oversight of underground injection disposal projects, saying the division had been too lenient about how wide a radius well operators must study before injecting and how much pressure should be applied.

California voters approved a measure in 2013 to better regulate the practices, requiring the state to study the risks of both fracking and acidizing, and adopt regulations and a permit system. The new law also required that neighbors near projects be notified, public disclosure of the chemicals being used, and groundwater and air quality monitoring. In 2014, voters in San Benito and Mendocino Counties approved bans on fracking, while a similar measure was defeated in Santa Barbara County. The Los Angeles City Council took a first step toward a ban on fracking within the city limits, voting in February 2014 for a moratorium until the practice is proven safe. The moratorium stalled, however, as city planners sought expert advice.

Incredibly, as California struggles with water issues related to oil and gas, the state remains the only government in the world not to charge any severance tax for extracting oil deposits, though each county can access property taxes. Daily oil production in 2012 was 541,000 barrels, from over 200,000 wells, and California ranks fourth among oil producing states. Yet the state gets no direct tax benefit from extraction of its oil reserves and has not established a fund to clean up environmental damage caused by its extraction, production, or transport.

CAN YOU DRINK THE WATER?

Clean water, like clean air, is our birthright, affirmed on September 2012, when the Human Right to Water Act became law, committing California to "ensuring affordable, accessible, acceptable and safe water sufficient to protect the health and dignity of all its residents." That was also the point of the CWA enacted four decades earlier, in 1972, passed "to restore and maintain the chemical, physical, and biological integrity of the nation's waters." States were required to develop lists of impaired waters, rank those waters, and develop TMDLs for each source. The early focus was on easily identifiable point sources, such as large factories and power plants. Nonpoint sources, the millions of impacts on water quality that are integral to our daily life, were a more intractable challenge.

The scope of the nonpoint-source problem became apparent when the U.S. Geological Survey (USGS) issued the results of a 2002 study of 139 streams in 30 states, including 10 in California. In those streams, they found a soup of products that are very familiar to the 39 million residents of this state, including antibiotics, ibuprofen and other painkillers, antacids, codeine, contraceptive hormones, perfumes, fire retardants, and caffeine. The USGS found the Sacramento River, just south of the capital city, contained "higher levels of acetaminophen (the active ingredient in Tylenol), birth-control hormones and cholesterol than most waterways tested" across the nation. The message in these findings for human health, and for the rest of the life forms dependent on water habitats, is not yet known.

Another message is being sent from our beaches. Over 150 million visits by tourists and residents spend over $10 billion each year in California. In 2001, from Memorial Day through Labor Day, 613 health warnings were posted to close Orange County beaches and harbors. Bacteria levels were too high for swimmers and surfers. Those that ignored the warnings risked gastrointestinal problems; eye, ear, nose, and throat infections; and even hepatitis (fig. 91). In 2003, over 1,500 miles of beach were temporarily closed. Closures happen following sewage spills or increased runoff after major storm events and floods.

Part of the problem is the number of people living in the coastal watersheds. Another part is that 80 percent of the urban landscape has impervious surfaces, like buildings and roads, so that rainwater instantly runs off instead of settling into the

Figure 91. Closure notice at the Doheny State Beach lagoon in summer 2001.

ground. Most runoff entering storm drains is never treated. It carries pollutants from suburban yards into gutters and drains and then, typically, into water channels draining to the sea. Though San Francisco Bay is ringed by heavy industry, up to 70 percent of the pollutants entering the bay may originate from city streets and gardens. Urban dwellers have become familiar with stenciled messages like "Drains to the Ocean" on sidewalks above gutters, and with information campaigns that tell them, "You Are the Solution to Stormwater Pollution." Those are cautionary reminders about our connections to local watersheds and the greater water cycle (fig. 92).

Waste that actually receives some treatment has historically been responsible for part of the problem along Orange County's

Figure 92. "Drains to Bay" message discouraging dumping of oil and other pollutants into the storm drain.

beaches. After passage of the federal CWA in 1972, sanitation districts in Orange County and San Diego had been granted waivers allowing them to pump sewage several miles offshore after only primary treatment. Evidence grew that a massive plume of contaminated water accumulated at the Orange County sewage outfall and may have circulated back toward the beaches. The working theory had been that cold ocean water would "cap" the warmer layer where the sewage was being dumped, holding the plume in place. But studies by the University of California at Irvine and Scripps Institute of Oceanography suggested that circulation patterns were more complex and could possibly return sewage toward the shore. Christopher Evans, executive director of the Surfrider Foundation, made a good point in a *Los Angeles Times* story (May 17, 2002): "To argue about whether we should be dumping 240 million gallons of partially treated sewage in the ocean...is goofy in the extreme."

The Board of Directors of Orange County Sanitation District (OCSD) voted in 2002 to drop their waiver and proceed with facilities providing full secondary treatment. A large percentage of Orange County's wastewater has, since 2012, been further treated by microfiltration, reverse osmosis, and advanced oxidation in Orange County Water District's groundwater replenishment system.

In 2009, San Diego was granted another five-year waiver, because monitoring by the Scripps Institute of Oceanography indicated no significant ecological effects from the ocean discharge. The city is embarking on a "Pure Water San Diego" water recycling program that will reduce ocean discharges (see the "Recycle and Reuse: Localizing Water" section, Chapter 5).

Water agencies generally follow six steps to clean their water supply before sending it to domestic users: primary treatment involves aeration, spraying as in a fountain to release gases and coagulation, to cause fine particles to join together, followed by flocculation, which mixes sediments so they combine and settle out. Secondary treatments take advantage of biological processes during sedimentation in quiet basins or constructed wetland systems, which can remove about 85 percent of suspended matter. Tertiary treatment is any further processing, including filtration through coal, sand, or gravel, then disinfection.

Small amounts of poisons are added to our water to kill microbes and control infectious diseases. For decades, the most common disinfectant used was chlorine. Chlorination saved millions of lives by reducing the incidence of typhoid fever, cholera, and other diseases spread by water. Unfortunately, it can also produce carcinogenic and mutagenic by-products. Water providers seek to assess and balance these risks. To control human exposure, the EPA sets standards for disinfectants and their by-products, and the SWRCB and Regional Water Quality Control Boards enforce them. The SWRCB is responsible for the regulatory oversight of about 8,000 public water systems throughout the California.

Overexposure to chlorine compounds can cause eye, nose, and stomach discomfort (chlorine and chloramines); nervous system effects (chlorine and chlorite); and, in infants and young children, anemia (chlorine and chlorite). Cancer risks are associated with other disinfectant by-products, such as bromate, haloacetic acids, and trihalomethanes (THMs). THMs also cause liver, kidney, and central nervous system disorders. THMs form in chlorinated waters that contain high concentrations of dissolved organic compounds and bromide. This is a big issue in the Sacramento–San Joaquin Delta, where the estuary naturally generates lots of organic material, and seawater delivers the bromide.

A very large study by the California Department of Health Services found that the risk of spontaneous abortion is 80 percent higher among women who drank five or more glasses of chlorinated tap water with relatively high THM levels during pregnancy than among women with lower exposures.

Approximately 30 percent of U.S. drinking water utilities have shifted to disinfecting with chloramines, containing chlorine and ammonia, which have fewer toxic by-products than chlorine alone. Over time, it was learned that chloramines can lead to some problem by-products too, including nitrosamines. The long-term answer can be a shift to alternate technologies, which also eliminate accidental chlorine spills during transport and handling. Slow sand filtration mimics the natural filtering of groundwater, creating an "ecosystem" of tiny organisms that consume organic matter and disease-causing microbes. Ultraviolet radiation controls microbes by damaging their DNA and is cheaper than chlorine. Ozonation takes advantage of a powerful oxidant, ozone, that breaks down bacteria, viruses, pathogens, and organic compounds. It is expensive and also can produce some THMs. EBMUD uses ozonation at two of its water-processing plants, however, and has shown that, as long as an adequate dose of ozone is supplied, ozonation causes no measurable increase in the mutagenic activity of water. A problem with each of the nonchemical techniques is that water travels away from treatment locations without carrying a disinfectant along into the delivery pipes, and might be recontaminated before it is used.

The Bottled-Water Phenomenon

Public concerns about tap water quality are partly responsible for the billion-dollar bottled-water industry in the United States. People are willing to spend from 240 to over 10,000 times more per gallon for bottled water than they typically do for tap water. At supermarkets, the cheapest bottled water sells for about a penny an ounce. The great majority is sold at two or three cents per ounce. Fancy imported bottles of water cost up to nine or 10 cents per ounce. At the average price range, one gallon of water costs between $2.90 and $5.30, and each AF grosses from $945,000 to $1.7 million for retailers! By comparison, the MWD charges about $400 per AF. Are buyers getting their money's worth? In particular, are they buying water that is purer than what comes out of their taps? (fig. 93).

Almost universally, labels on bottled water picture mountain snow, glaciers, or springs that emerge from the ground in a pristine, natural setting. The messages are of "purity" that comes from "nature." But bottled water is not necessarily cleaner or

Figure 93. An entire row of supermarket shelves devoted to bottled water, a cultural phenomenon worth questioning.

safer than most tap water. The Natural Resources Defense Council (NRDC) tested more than 1,000 bottles of 103 brands of bottled water for contaminants, and summarized its findings in a March 1999 petition to Congress asking the Food and Drug Administration (FDA) to better control the industry. About one-third of the waters tested contained levels of contamination that exceeded allowable limits.

Bottled water has to be tested, but it is sampled less frequently than city tap water for bacteria and chemical contaminants. Current standards for bottled water still allow some (minute) contamination by *Escherichia coli* or fecal coliform, although *no* such contamination is allowable in tap water. And unlike tap water, bottled water is not required to be disinfected or tested for *Cryptosporidium* or *Giardia* parasites.

Amazingly, the NRDC found that about one-fourth of bottled water was nothing but tap water. Labels that read "from a municipal source" or "from a community water system" indicate that the water, however scenic its label, originated as tap water. Some, but not all, of those brands provide additional filtering or treatment. FDA rules allow bottlers to call their products "spring water," suggesting a natural source emerging from the ground, even though the water is extracted with pumps and then treated with disinfectants. The NRDC reported one label that read "spring water," with a picture of a lake surrounded by mountains, on bottled water that originated from an industrial parking lot next to a hazardous waste site.

The bottled-water phenomenon is an expensive and disturbing reaction by consumers, despite repeated assurances from municipal water providers that their products are safe and clean. Recently the MWD discovered that Central American immigrants were the state's biggest consumers of bottled water. Raised in lands where tap water was dangerous, they did not trust domestic drinking water sources here. Despite our potable water standards and higher level of treatment, the attraction of bottled water may primarily be convenience. The smell of chlorine in tap water (which will go away soon if the water is stored in a pitcher or canteen), may explain some people's preference for bottled water.

The greenhouse gases generated because of the national craze for bottled water are another concern about an industry that has grown by nine percent a year since this book first explored the "bottled-water phenomenon." Manufacturing 28 million water bottles a year for US sales requires 17 million barrels of oil, and then even more fossil fuels as the bottles are shipped across the nation or imported from other countries. Making plastic, one-time-use bottles consumes three times as much water as goes inside each container and releases 2.5 million tons of carbon dioxide to the atmosphere each year. A number of California cities, including San Francisco, Los Angeles, and San Jose, finally reacted to these environmental costs by canceling municipal contracts for bottled water in their offices and at public events. "Ban the Bottle" campaigns at UC Berkeley, UC Santa Cruz, Humboldt State University, and other colleges and high schools have convinced schools to cancel contracts with bottled-water vendors on campus.

Mass Medication

Community water fluoridation has been endorsed by the American Dental Association, the American Medical Association, the American Association of Public Health, the U.S. Public Health Service, the Centers for Disease Control (CDC), and the World Health Organization. Since 1997, California state law has required fluoridation by all communities with populations greater than 10,000 (so long as they have outside funding, rather than tax money, to install and operate the systems). About one-third of Californians drink fluoridated water. With such expert and legal backing, it might seem sensible to medicate all of the population this way. Yet voters in Cotati, Crescent City, Davis, Hoopa Valley, La Mesa, Los Altos Hills, Mammoth Lakes, Modesto, Napa, Olivehurst, Redding, Santa Barbara, Santa Cruz, Suison City, Watsonville, and Woodside have rejected proposals to fluoridate their water supplies. What is a common citizen to think about vocal opposition whenever a community considers this step?

Since the 1950s, each generation has been brought up hearing commercials for fluoride toothpastes that told them the chemical was good for them. Yet the fine print on tubes covered with cartoon characters reads: "WARNING: As with all fluoride toothpastes, keep out of the reach of children under six years of age. If you accidentally swallow more than used for brushing, seek professional assistance or contact a Poison Control Center immediately." Too much fluoride intake can rob bones of calcium, making them weak and brittle (skeletal fluorosis). During the tooth-forming years (age 8 and younger), dental fluorosis can produce tooth enamel with faint white mottling. Children older than eight years, adolescents, and adults are not susceptible to dental fluorosis. Cautionary messages warn doctors and parents not to give fluoride supplements to infants under the age of three or to pregnant women. Infants in the first year of life, exclusively fed with formula mixed with fluoridated water, have an elevated risk for fluorosis. Parents can instead use low-fluoride bottled water to mix infant formula, labeled as deionized, purified, demineralized, or distilled.

Part of this debate has centered on the argument that direct applications to teeth by dentists are more effective than drinking water containing fluoride. Questions about efficiency and unintentional effects also arise. Only seven percent of residential water is used for drinking, so 93 percent of the fluoride never "sees" a tooth before it washes down some drain to cycle back to the environment. Fluoride accidentally released in high concentrations into rivers has caused delayed salmon migrations, deformed embryos in salmon and trout, and accelerated mortality rates. The sodium fluoride used in water supplies is an EPA-listed hazardous waste product of the phosphate fertilizer industry. In concentrations far higher than those allowed in drinking water, it is poisonous and is used to kill insects and rats. It is the toxic element in the nerve gas Sarin.

Of course, many helpful substances are toxic at high concentrations. The control of overall exposure is the key to safe use. In 2011, the U.S. Department of Health and Human Services Agency recommended that water systems adjust their fluoride content to 0.7 parts per million, the low range of previous recommendations that had been as high as 1.2 parts per million, to minimize the chance of children developing dental fluorosis. The standard fluoride dose is based on estimates of average daily water consumption, but today Americans are exposed to fluoride from many sources. Besides toothpastes and mouth rinses, prescription fluoride supplements, and fluoride applied by dentists, it can be present in bottled juice and beverages and other products made or processed with fluoridated water. The challenge of fluoride dosage control is at the heart of this debate. The SWRCB Division of Drinking Water publishes annual reports on monthly fluoride levels in the state's public water systems.

Domestic water systems in Europe remain 98 percent fluoride free. Low cavity rates there are apparently related to high standards of living, less eating of refined sugar, regular dental checkups, flossing, and frequent brushing.

Dentists, health providers, and lawmakers who promote fluoridation are obviously well intentioned. Some argue that the water supply is the only way that the poor will receive fluoride applications. This is a debate that revolves around society's willingness to solve one problem without creating others along the way.

Giardia

One of the greatest pleasures of hiking in the Sierra Nevada used to be drinking directly from the clean mountain streams. Then, in the final decades of the twentieth century, came warnings

Figure 94. Filtering water in
the backcountry to minimize
the risk of *Giardia*.

about *Giardia lamblia*, apparently a threat in every stream and lake
in the mountain range. Backpackers began boiling their water,
lugging heavy bottles along the trails, or carrying special filter
pumps or iodine disinfectants (fig. 94).

Giardia is the most commonly diagnosed intestinal parasite in
North America. It is the most frequently identified cause of
diarrheal outbreaks associated with drinking water in this country; the CDC estimates that as many as 2.5 million cases occur
annually. Studies have shown that 10 or so cysts are required to
create a reasonable probability of contracting giardiasis; about
one-third of persons ingesting 10 to 25 cysts pass detectable cysts
in their stools.

In 1984, the USGS and the California Department of Public
Health examined 69 Sierra Nevada streams, two-thirds of which

were considered to have a high probability of human fecal contamination. *Giardia* cysts were found at 43 percent of the high-use sites and at 19 percent of the low-use sites, but always at very low concentrations. The highest concentration was 0.108 cysts per liter of water in Susie Lake, south of Lake Tahoe. The next highest was 0.037 per liter near Long Lake, southwest of Bishop. Samples taken in the Mount Whitney area varied from zero, at most sites, to 0.013 in Lone Pine Creek.

By comparison, San Francisco city water, renowned for cleanliness that allows it to be delivered unfiltered, can contain 0.12 cysts per liter, a higher concentration than that measured anywhere in the Sierra. Los Angeles Aqueduct water, with only 0.03 cysts per liter, has a higher concentration than all but two of the sites surveyed in the Sierra.

Robert Rockwell, who summarized these data on *Giardia* for a 2002 article that questioned widespread concerns about backcountry water, concluded that good backpacker hygiene combined with common sense choices about where to pull drinking water from streams (upstream from trails, away from camps, in good flows with plenty of aeration) may be far more important in avoiding giardiasis than treating or filtering water. "Better safe than sorry" may be the philosophy that dictates water treatment, despite these findings.

Where Does Your Dog Go?

You may diligently control your trash and oil and pesticide use. Do you as diligently control your dog? We would never think to defecate on the ground, wherever we might be when the need arises, but many of us allow our pets to do so without giving any thought to the consequences. Most people pick up after their pets in their own yards, but many dogs roam (despite leash laws), and many dogs are taken for walks specifically to move the problem out of backyards. Then there are outdoor cats, tidily burying wastes out of sight. The issue of pet sanitation and polluted water has remained invisible to most people, another example of how Californians ignore the impacts of their own unending population growth. As the human population increases, so do the numbers of our "fellow travelers," beloved dogs, cats, and other pets, all generating wastes every day.

The American Pet Products Manufacturers Association's annual National Pet Owners Survey estimates that four in 10 (or 40 million) US households own at least one dog. The average owner has about two dogs. With about 39 million people in California, a formula developed for veterinarians to estimate market statistics generates an estimate of 7.9 million dogs and 8.8 million cats in the state. They are concentrated where their human owners are concentrated, in the urban centers, particularly along the Southern California and San Francisco Bay coastal areas.

As much as 95 percent of the fecal coliform in urban stormwater, in some studies, had a nonhuman origin. A Seattle study found that nearly 20 percent of the stormwater bacteria came from dogs. Dog feces can be a significant source of fecal coliform and fecal strep bacteria. Dogs are also hosts of *Giardia* and salmonella. Another study, published in July 2002, suggested that runoff to the ocean may be introducing a parasitic infection, toxoplasmosis, responsible for the deaths of sea otters. Domestic cats are the only animal group known to pass along the eggs of that parasite.

Proper cleanup and disposal are essential habits. The street gutter is the wrong place for feces disposal. Pet population control is another important part of the equation (as is human population stabilization, for that matter).

THE PROBLEM IS US

Population growth continues to exacerbate California's water supply problems. Growth projections have been used to justify water development projects ever since the first California Water Plan was prepared in 1957. When used to justify development, such population estimates become self-fulfilling prophecies. Predictions about population numbers, climate change, and water supply are not inevitable outcomes, though our society has a habit of extending graph lines outward and proclaiming *that* vision of the future as certain.

"Economic growth, of course, depends on population growth," a *San Francisco Chronicle* editorial told its readers on August 6, 1989. "Population growth depends absolutely on guaranteed – and continuing and growing – supplies of good quality drinking water." That was the dogma that shaped the twentieth century history of California. There is almost no room within that belief system for contemplation of long-term sustainability, carrying capacity, or optimal growth limits. Yet, growth rates have slowed significantly in California and are at sustainable rates in many other states and developed nations. Accepting the editorialist's dogma could keep us from actively pursuing a sustainable future. Water limits are real. Though there are significant efficiencies possible through conservation, recycling, and groundwater management, those measures will, ultimately, be overcome by never-ending population growth.

One economic pressure on water consumption arrived with the computer industry's growth in Silicon Valley and the regional population growth it engendered (fig. 95). The concentration of high-tech industry at the south end of San Francisco Bay could also be called "Thirsty Dotcom." Computer memory

Figure 95. Santa Clara Valley, once a major agricultural region and now better known as Silicon Valley, its growth made possible by imported water.

chips are made by etching circuit patterns, using acids and solvents, onto silicon wafers, which are then cut into many chips. Thousands of gallons of ultrapure water are used to rinse off the chemicals. Washing a single eight-inch silicon wafer uses about 2,000 gallons of water. Silicon Valley electronics firms collectively use millions of gallons of water each day. Serving that thirst has added to the pressures on the Sacramento–San Joaquin Delta.

The water-for-growth paradigm is bumping against limits in the twenty-first century, with conflicts between urban and agricultural demand and severe environmental impacts. We face a daunting list of challenges. Yet the future is in our hands.

CHAPTER FIVE

Meeting the Challenges

California's Water Future

The general welfare requires that...the waste or unreasonable use...of water be prevented, and that the conservation of such waters is to be exercised with a view to the reasonable and beneficial use thereof in the interest of the people and for the public welfare.

    —California State Constitution, Article X, Sec. 2

The public trust...is an affirmation of the duty of the state to protect the people's common heritage of streams, lakes, marshlands and tidelands.

    —Supreme Court of California (1983)

Every human being has the right to safe, clean, affordable, and accessible water adequate for human consumption, cooking, and sanitary purposes.

    —California Human Right to Water Act, 2013

CALIFORNIA WATER LAW
AND THE PUBLIC TRUST

The future of California water management will be guided by our laws. All water within California is owned by the state on behalf of the people. Since 1928, the California Constitution has required that water be put to the highest beneficial use, which was interpreted, through most of the twentieth century, as domestic or profit-making uses. The constitution also prohibits waste or unreasonable uses.

One of the first legal decisions to regulate impacts from water use came in 1884, when hydraulic miners were ordered to keep their waste-laden runoff confined on their own property so it would no longer damage property owners downstream. The state's current water rights system traces back a century to the Water Commission Act of 1914 which formalized two kinds of water rights: riparian and appropriative. Riparian water rights come with land ownership adjacent to a river or lake and are "senior" in priority to appropriative rights. Moving water away from its source for use elsewhere requires appropriative rights. A water right is either "pre-1914" or "post-1914": those granted before the 1914 dividing line have senior priority and post-1914 have junior status. Beyond that dividing point, priority is a matter of "first in time, first in line." When water becomes scarce, those distinctions determine who has to stop using water first.

In 1931, as a reaction to Los Angeles taking of Owens Valley water, "county of origin" provisions were written into the California Water Code to assure that water exports never deprived the county where the precipitation fell enough water for its future development. Later, as the federal Central Valley Project (CVP) began, "watershed of origin" provisions were added to protect the "beneficial needs of the watershed, area, or any of the inhabitants or property owners therein" as first priority, ahead of water export projects. In 1937, the Fish and Game Law to protect fisheries in streams below dams or stream diversions was enacted.

In the 1970s, several major environmental protection laws were passed in reaction to degradation and losses that had been accelerating since World War II. In that decade, the federal Clean Water Act and Safe Drinking Water Act, the federal and state Endangered Species Acts, and the federal and state Wild and Scenic Rivers systems were created. In 1983, a significant court decision, that the Public Trust Doctrine applied to Mono Lake, shook up long-established assumptions about water rights, water law, and the environment in California. Another important change in water law came in 1992 with the CVP Improvement Act, which returned 800,000 AF of annual runoff to the environment. In 1993, the California Wetlands Conservation Policy was adopted, aiming for no net loss of wetlands in the state. Today, 28 percent of the state's water is dedicated by law to environmental uses in wetlands, in the San Francisco Bay and Sacramento–San Joaquin Delta, and in state and federally designated Wild and Scenic Rivers.

The second decade of the twenty-first century brought a flurry of water laws, regulations, and planning addressing challenges that are emerging as water limits bump against decades of population growth and development. The Water Conservation Act of 2009, better known as "*20×2020*," set a goal to reduce per capita urban water use by 20 percent by December 31, 2020 (California Department of Water Resources 2010). Since September 2012, when the Human Right to Water Act became law, relevant state agencies must consider that fundamental birthright as they carry out their duties. The Rainwater Capture Act of 2012 made it legal for individuals to divert rainwater from roof gutters into cisterns for watering yards, and the California Plumbing Code was modified in 2013 to allow graywater systems to divert drainage from washing machines and showers toward landscape use. In 2014, the governor's California Water Action Plan laid out key actions for the following five years toward sustainable water management.

The State of California's constitutional provisions against unreasonable use of water and the court decision that the Public Trust Doctrine applies to Mono Lake have particularly broad significance. The "reasonable use" requirement in the state constitution, itself, reflects the Public Trust doctrine, which recognizes that government has a legal responsibility to protect resources with environmental and aesthetic values. Such resources are held "in trust" for the public. The public trust covers the people's rights to use California's water resources for navigation, fisheries, commerce, environmental preservation, and recreation; as ecological units for scientific study; as open space; as environments that provide food and habitat for birds and marine life; and as environments favorably affecting the scenery and climate of the area. In the 1983 decision, the court also identified "changing public needs" for resource protection as requiring action by the government, adding, "Thus, the Public Trust is more than an affirmation of state power to use public property for public purposes. It is an affirmation of the duty of the state to protect the people's common heritage of streams, lakes, marshlands and tidelands, surrendering that right of protection only in rare cases when the abandonment of the right is consistent with the purposes of the Trust."

Application of the Public Trust Doctrine to Mono Lake rattled complacent water agencies in California. The city of Los Angeles had held legal water licenses authorizing its diversions from Mono Basin streams for almost a half century. Despite that length of time, the court held that such water rights could be reevaluated in

Figure 96. Phalaropes at Mono Lake. Well over one million migratory and nesting birds visit the lake each year.

the light of new information and changed societal values. Protection for the environment, however, was not absolute. Trust values had to be *balanced* with beneficial domestic uses of diverted water. Los Angeles did not stop diverting water out of the Mono Basin, but it was required to reduce its take to levels that would also protect the Mono Lake ecosystem and its tributary streams.

Since 1941, four major streams that fed Mono Lake had been diverted into the Los Angeles Aqueduct. Their complete dewatering also violated the state Fish and Game Code, and the court's decision corrected that violation as well. Without the stream water to replace evaporation, Mono Lake had lost half of its volume and the salinity of its already harsh alkaline waters doubled. The unusually productive ecosystem, with its brine shrimp and Alkali Flies *(Ephydra hians)* that fed over a million migratory birds, was in danger of collapse. Island nesting habitat for the biggest breeding colony of California Gulls *(Larus californicus)* in the state had become connected to shore and was invaded by coyotes as the lake dropped 45 vertical feet (fig. 96).

The Mono Lake lawsuits were initiated by the National Audubon Society and the Mono Lake Committee, with CalTrout later suing separately over the dewatering of the streams. Under a plan ordered by the State Water Resources Control Board (SWRCB) in 1994, the lake is being refilled to an elevation that still will be 27 feet below its prediversion level (a compromise that also recognizes beneficial uses of water diverted to Los Angeles). The recovery period, it was estimated in 1994, would last about 20 years, but in 2015 the lake still was more than 10 feet below the target elevation. A series of drought years and on-going climate drying were interfering with the societal desire to correct a half century of stream diversions to the lake.

Nevertheless, "Save Mono Lake" messages on bumper stickers have been replaced by "Long Live Mono Lake" and "Keep Saving Mono Lake" messages. Active stream and waterfowl habitat restoration measures are underway, recognizing that it could take centuries for the damaged streambeds to recover naturally and that lower lake levels will not ever restore original conditions for millions of ducks and geese that once used the lake.

With the legal precedent of the 1983 Public Trust Decision and its affirmation by review courts, the doctrine's requirements now are incorporated in each decision made by state agencies managing water resources.

THE DELTA, A TUNNEL VISION, AND A WATER BOND

Dire "water crisis" pronouncements were heard in the early years of the twenty-first century as less water became available via all the major aqueducts and courts ordered diversions from the Sacramento–San Joaquin Delta to be temporarily reduced. Severe declines were occurring in smelt, shad, and salmon species that rely on flows through the estuary and out to sea. Possible causes included water pollution and competition from introduced species, but years of increased pumping into the southward flowing aqueducts coincided, tellingly, with the population declines. From 1990 to 1999, the average amount of water pumped away from the Delta had been 4.6 MAF, but from 2000 to 2007, the State Water Project (SWP) and CVP extractions increased to 6 MAF. The highest yearly totals came in 2003 (6.3 MAF), 2004 (6.1 MAF), 2005 (6.5 MAF), and 2006 (6.3 MAF).

In March 2007, a state court ruled that the Department of Water Resources (DWR) was violating the California Endangered Species Act by not protecting threatened Delta Smelt *(Hypomesus transpacificus)* and endangered salmon species from the impacts of its operations. Pumps serving the California Aqueduct were shut down for 10 days. When only 25 Delta Smelt could be found in the annual population counts in May 2007, a federal court ordered 30 percent reductions in both state and federal pumping from that December through June 2008, the months when smelt would mass near the pumps. The U.S. Fish and Wildlife Service and National Marine Fisheries Service were ordered to come up with effective protection plans for the endangered species.

The 2008 salmon fishing season was canceled because the fall run of Chinook in the Sacramento River had drastically declined. Another Delta fish species, the Longfin Smelt *(Spirinichus thaleichtys)*, was also being considered for listing under the Endangered Species Act. Researchers were documenting problems all the way down the food chain to plankton, the primary producers that support other life.

The governor appointed a seven-member Delta Vision Blue Ribbon Task Force in 2006 to develop goals for sustainable management of the Delta, considering environmental quality and economic and social well-being. The task force issued a 12-point list of recommendations in January 2008. They identified "coequal goals" of (1) a sustainable ecosystem and (2) a reliable water supply, and recommended a significant increase in statewide conservation and water system efficiency efforts, new facilities to store and convey water, and likely reductions in amounts of water taken out of the Delta watershed.

The Task Force also recommended there be a new governing entity for the Delta and so, in 2010, the Delta Stewardship Council was established as an independent state agency. On September 1, 2013, the Delta Plan, a comprehensive, long-term management plan, was adopted with legally enforceable regulations guided by the coequal goals. Under the legislation that created the Delta Stewardship Council, the CALFED Science Program became the Delta Science Program.

CALFED had been an ambitious attempt, begun in 1994 by the state and federal governments, to do something quixotic and perhaps impossible: end California's long history of water wars. No one (except, perhaps, certain lawyers) wanted to face the costs and time required for complicated water litigation. The Bay-Delta Accord outlined basic principles and water quality standards to guide the program. Twenty-three state and federal agencies and many public and private groups participated in CALFED. They sought balanced solutions through consensus among "stakeholders," generally categorized as "urban interests," "agriculture," and "environmental advocates." The three stakeholder categories may not accurately reflect the real range of values. "Agricultural interests" include farmers and related agribusinesses, but also many of

California's farm corporations that, history shows, are as interested in real estate development as in long-term agriculture. "Environmental organizations" have memberships that are overwhelmingly urban, yet their goals are generally very different from those of the stakeholders labeled "urban interests." That group includes public service water providers, but also interests that focus heavily on overcoming water limits that would curtail growth and development. Establishing consensus allowed the process to begin, but it ultimately became impossible for CALFED to do everything for everyone.

Projects were funded on more than 70 percent of California's landscape, across 40,000 square miles of watershed lands. Many restoration efforts focused on the Sacramento–San Joaquin watershed, including fish passage improvements along 182 river-miles and 64 new fish screens at diversion points. Water users were motivated to correct environmental conditions as a way to improve both the quality of the water coming to them from the Delta and its reliability, so that pumping from the Delta might not so often be curtailed.

Several issues hindered CALFED, including disagreement on what water uses should be prioritized, ways to evaluate outcomes, and uncertain financing (the operations budget relied upon state, federal, regional, and local water agency funding), and the formal effort ended in 2009. However, some elements of CALFED continue, including a 30-year water use efficiency grants program and useful studies evaluating new reservoirs options. The effort that took the place of CALFED was called the Bay Delta Conservation Plan (BDCP).

The BDCP proposed the largest and most expensive water infrastructure project ever in California, in fact, the most expensive public works project in US history. It aimed to deliver

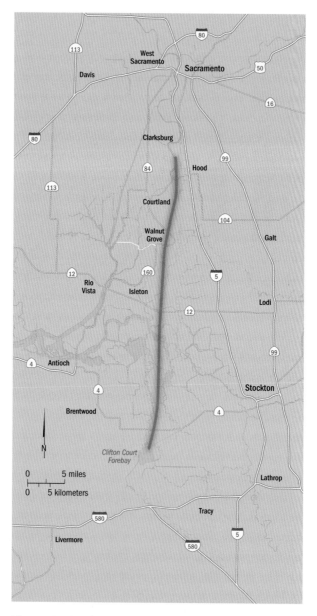

Map 27. Proposed route of twin tunnels to carry Sacramento
River water beneath the Delta, part of the California Water
Fix plan.

Figure 97. In 1982, when a peripheral canal proposal was defeated in a statewide referendum, this *San Francisco Examiner* cartoon predicted the canal's resurrection. In 2014, the idea was back as twin tunnels beneath the Sacramento–San Joaquin Delta.

Sacramento River water to the south Delta pumps in two massive underground tunnels, each 40 feet wide and 35 miles long (map 27). The twin tunnels proposal was the latest alternative to the peripheral canal that was defeated in a statewide referendum vote in 1982 (fig. 97). Throughout eight years of planning, still ongoing in 2015, BDCP proposals remained very controversial. More narrowly focused than the Delta Plan, it was a "habitat conservation plan" under the US Endangered Species Act and, under California law, a "natural community conservation plan."

To address conservation, the BDCP proposed to gradually return about 153,000 acres of the Delta to slow-moving sloughs, riparian forest, grassland, and tidal marsh habitat lost during the prior 150 years to farms, levees, roads, and human development. Conditions for 57 species of endangered fish and other animals in the estuary were predicted to improve as the tunnels provided an alternative to pumping water from the south edge of the Delta. When the powerful aqueduct pumps operate, they pull Delta water toward them, reversing flows that naturally moved east to west toward the ocean. That confuses migrating salmon trying to navigate upstream and kills Delta Smelt when they are pulled into the pumps. Though the south Delta pumps will not stop being used, about half of the diversions would shift to the tunnels. Fewer fish should die, so there should be less likelihood of mandated shutdowns. That is the primary way the plan aimed at improved reliability for SWP and CVP water customers south of the Delta. The project was, also, a climate change adaptation tactic for dealing with predicted impacts to levees and from salinity intrusion due to rising sea level.

The tunnels would be bored 150 feet below the Delta surface and convey water without pumps by gravity flow. Engineers are confident that existing technology can successfully bore the tunnels. Initial costs are projected to be $24.5 billion, with water districts and utilities that use SWP and CVP water passing the costs of design, construction, operation, mitigation, and adaptive management to their customers. Another $7 billion is needed for habitat restoration, pollution control, anti-poaching and other ecological programs, and would be funded by taxpayers through state and federal agencies. The total cost during the 50-year project will exceed $54 billion, factoring $26.3 billion in interest on tunnel revenue bonds.

The administration moved ahead on the BDCP without legislative action or voter approval (although the legislature may be asked to appropriate parts of the funding). According to Governor Jerry

Brown, the 1960 bond act that authorizes the SWP under Governor Pat Brown's leadership also allows for the construction of "Delta facilities," and that grants the administration this authority.

The plan called for no *increase* in diversions out of the Delta. The maximum capacity proposed for the two tunnels, together, was 9,000 cubic feet per second (cfs). That was big enough to capture the entire flow of the Sacramento River during dry periods, but the coequal goal of a healthy estuary means that complete diversion must not occur. The most critical time for Delta fish is December through June. Proposed operation criteria for the tunnels would allow diversions at full capacity only when the river flows exceed 30,571 cfs, would have diversions at no more than 384 cfs when the river is running below 6,400 cfs, and stop all tunnel diversions should the river drop to less than 5,000 cfs.

Yet, concerns that were expressed during the 1982 Peripheral Canal referendum still persisted, that, once tunnels are in place with so much capacity, powerful interests might be able to sway tunnel operations to take *more* water from the Delta, to serve growing thirst for farm irrigation in the San Joaquin Valley and never-ending population growth in Southern California. Otherwise, leery opponents asked, why build on such a scale?

On the other hand, some water users south of the Delta were concerned about the cost–benefit balance of a project that actually intends to send less water their way. Because construction funding must come from water users, rather than from state or federal coffers, the cost–benefit analysis had to "pencil out" for those water districts. The California Legislative Analyst's office, in a January 2015 report, reminded readers that most major public infrastructure projects cost a third more than their original budget, and also warned of uncertain funding as water users may be unable or unwilling to continue to fund the massive project. Tehachapi-Cummings County Water District, one of those that provided initial planning money, declined to provide additional funding when requested in 2015, reluctant to proceed unless *more* water exports will be the result.

The BDCP calculated economic benefits under a low-outflow (toward the ocean) scenario (a high figure of 5.59 MAF per year exported by aqueducts) and a high-outflow scenario (less water to the aqueducts at 4.705 MAF, due to additional water sent to the ocean in the spring). The plan concluded that net benefits were $5.3 billion and $4.5 billion, respectively. But economist, Dr. Jeffrey Michael, Director of the Business Forecasting Center at the University of the Pacific, did an independent analysis in 2014 that concluded the BDCP would cost $2.50 for every $1 in benefits. These types of technical debates pose a great puzzle for concerned citizens.

The ecosystem restoration portion of the project will be funded by state and federal agencies. Though a water bond approved by voters in 2014 allocated $140 million for restoration projects, that was not intended for use on the BDCP project. During the bond campaign, the promise was that Proposition 1 would be "tunnel neutral," and not further the separate effort. The total cost of habitat and conservation in the BDCP is much higher, anyway, at $7 billion.

The complex details in this deliberation have too often been characterized as "fish versus farmers." Saving the tiny Delta Smelt is regularly criticized by opponents, ignoring the declining species' role as a "canary in the coal mine" indicator of ecosystem health overall. The flowing estuary is "salmon water" too. Farmers on islands within the Delta itself are at odds with San Joaquin Valley farmers, seeing the livelihoods threatened should less water flow through the estuary.

There have been alternative plans suggested that were not included in the BDCP environmental review documents. A coalition of environmental and business groups and some urban water agencies made a detailed proposal in 2013, for a single, smaller tunnel and fewer acres of habitat restoration, using money saved for levee improvements and water saving efforts south of the Delta, including water recycling and surface or groundwater storage. The diverse coalition included the Natural Resources Defense Council (NRDC), Defenders of Wildlife, The Bay Institute, Planning and Conservation League, Environmental Entrepreneurs, Contra Costa Council, the San Diego County Water Authority, Contra Costa Water District, the City of San Diego, E.B.M.U.D., the San Francisco P.U.C., Alameda County Water District, and Otay Water District.

An "elephant in the room," with influence over this debate, has been lurking ever since August 2010, when a Flow Criteria Report for fisheries protection was developed by the SWRCB. The report found that "the best available science suggests that current flows are insufficient to protect public trust resources. Restoring environmental variability in the Delta is fundamentally inconsistent with continuing to export large volumes of water from the Delta" (State Water Resources Control Board 2010). The flow criteria report did not consider any balancing of public trust resource protection with public interest needs for water. The conclusions were considered a radical threat to the status quo, in some quarters, and the discomforting report has mostly been treated as an "elephant" to ignore.

On April 30, 2015, state and federal agencies made a fundamental change to the BDCP, proposing to separate the tunnels from the conservation habitat elements, with a smaller restoration and protection effort on just 30,000 acres. This followed a critical review of the project's environmental documents by the US Environmental Protection Agency. The "Bay Delta Conservation Plan" was replaced by "California Eco Restore," the smaller habitat effort, and "California Water Fix," the twin tunnels project. The new proposal was to be evaluated in a recirculated draft EIR/EIS in 2015. With such a fundamental change following years of debate and conflicts about funding and objectives, the project's final outcome was not clear as this book went to print.

The Proposition 1 Water Bond (2014)

In June 2008, the governor declared California was officially experiencing a "statewide drought," after two drier-than-average winters. The Executive Order directed state agencies to improve water efficiency and facilitate water transfers. The governor also announced his support for an $11.5 billion bond negotiated by legislators, meant to be considered in the 2010 election. With poor chances of passage during an economic downturn, however, the bond was pushed back to 2012, and then pulled again from that election. After renegotiation reduced it to a $7.5 billion bond package, Proposition 1 was ultimately passed by voters in November 2014. The state was enduring a third year of, yet another, drought.

Bond funding for infrastructure included $810 million toward regional water conservation and rainwater capture, $725 million for water recycling and desalination, and $2.7 billion for dams or groundwater storage. There is another $1.5 billion for watershed protection and restoration, $1.4 billion to clean pollution in groundwater and protect against future contamination, and $395 million for repairing levees and flood protection. Projects will be considered for actual funding in a competitive process through the California Water Commission, which has nine members appointed by the governor. Important qualifications were placed on proposals that will guide and constrain the commission, prohibiting any projects that adversely affect a federal- or state-designated wild and scenic river. Any storage projects funded must also provide "measurable improvements to the Delta ecosystem or to the tributaries to the Delta," so are confined to that watershed.

The diverse categories funded reflect a shift away from dependence on the Sierra Nevada snowpack and on water imported via long-distance aqueducts, and toward increased reliance on local water sources, conservation, and better use of groundwater storage (fig. 98). This transition is driven by recent successes that paint an optimistic picture about California's water future. Despite a growing population, the state's total domestic water consumption actually dropped in recent years.

RECYCLE AND REUSE: LOCALIZING WATER

It seems insane to harvest water hundreds of miles away, pump it at a tremendous energy expense over mountains to southern California, clean and treat it before using it just one time, sanitize it again, and then dump it out to sea. Yet that has been the fate of more than 3.5 MAF of water each year in modern California. Consider how much water is present, as you read these words, in communities at the delivery end of water aqueducts, not just in reservoirs, but moving through underground supply and sewage pipes, to and from homes and businesses. And consider how much water is, right now, within the bodies of millions of people (and their pets) – millions of small "reservoirs" that

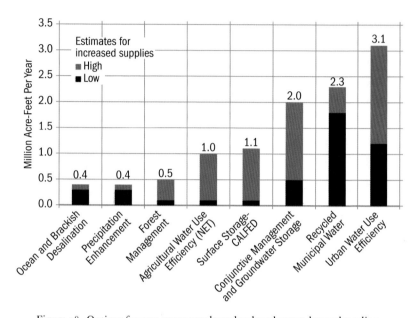

Figure 98. Options for new water supply and reduced water demand totaling 5 to 10 MAF per year. The greatest opportunities existed for conservation (urban water use efficiency), recycling of municipal water, and groundwater storage. Note how much less water was expected from dam construction (surface storage) or desalination (redrawn from Delta Stewardship Council 2013).

add up to amazing quantities of water. That urban water is in place, where it is needed, today. Through recycling, it can be reused again, and again, and again.

As water policy is debated, it has become commonplace to hear that "we can't conserve our way out of the water crisis." Those who make that statement may be behind the times. Successes in wastewater recycling in Orange County have served as models for other localities reusing water to replenish groundwater basins, irrigate landscapes, flush toilets, and supplement domestic supply reservoirs. In October 2014, the Mayor of

Los Angeles issued an executive directive to reduce water use by 20 percent by the year 2017 and cut imported water by 50 percent by 2024, by relying more on conservation, recycling wastewater, and cleaning polluted groundwater. The City of Long Beach and the West Basin Municipal Water District are aiming for 100 percent use, and reuse, of local water, thus weaning themselves completely from imported water.

Recycled water (sometimes called "reclaimed" water) is a drought-proof source of supply, because it is already "in hand" and affected very little by weather cycles. It does not have to be transported hundreds of miles, from mountain snowpacks that diminish during droughts, or taken from water-dependent habitats serving endangered species. Recycling, whether by nature or by humans, is one of the many "water wheels" that are shortcuts within the great planetary water cycle. It allows us to stop the folly of watering lawns and flushing toilets with highly treated drinking water.

Partially recycled sewage water has been applied to agricultural purposes for decades, but treatment levels have improved and types of uses are proliferating. The DWR documented 669,000 AF going to 11 categories of beneficial use in the 2013 State Water Plan. All but seven of the state's 58 counties had recycling projects. The recent emphasis has become "fit for purpose designer water" that matches treatments to uses. The most rigorous and costly processing is reserved for water that will be in contact with final food production and for "indirect potable reuse." Treated water receives a final cleansing through natural processes below ground or in surface reservoirs. Direct delivery into drinking-water systems was not yet happening in 2015, though cleansing systems had become so effective that the option was under study.

Figure 99. Purple pipes carry recycled wastewater treated to high standards at the West Basin Groundwater Replenishment District.

In 1991, the Irvine Ranch Water District (IRWD) became the first water district in the nation with health department approval for *interior* use of reclaimed water. High-rise offices run reclaimed water through separate toilet-flushing systems identified by purple pipes. Drinking water used within the specially plumbed buildings dropped by as much as 75 percent, because it was no longer being flushed away. Overall, 25 percent of the water in Irvine's system is now reclaimed, delivering more than 23.5 million gallons of recycled water per day. "Water is too valuable to be used just once," according to the IRWD.

Purple piping carries recycled water, treated to a level suitable for irrigation and industrial use but not for drinking (fig. 99). State law currently limits the uses of recycled water to concrete plants, snowmaking, freeway landscaping, commercial laundries, fountains, golf courses, schoolyard lawns, and crop irrigation. It is also used to recharge groundwater aquifers. There are now over 90 miles of purple pipe beneath the streets of San Diego, passing recycled wastewater to parks, road medians and golf courses. In the eastern part of Chula Vista, new developments are required to install purple pipelines.

Traditional wastewater treatments remove solids and add disinfectants. Recycling for potable use relies upon additional reverse osmosis filtering, microfiltration, and ultraviolet radiation, and yields a product akin to distilled water. Full treatment can remove even medical wastes and drugs not removed by traditional wastewater treatment. When injected underground, recycled water goes through natural filtration and further cleansing by soil organisms.

Reverse osmosis forces water through very fine pores of microfilter membranes. Typical drinking water carries fine particles in the 500 parts per million range. By reverse osmosis, the recycled water can exceed drinking water standards, taking fine particles down to just 50 parts per million in water, and then further dilute it with potable supplies. Orange County has new treatment plants aiming at three parts per *billion*.

Of course, *all* of Earth's water is recycled. "Wastewater" is just a temporary status until natural processes of the water cycle clean and distill the waste and return it as pure water. The 22 million Californians who drink water taken from the Delta should realize that their supply includes a portion that has passed through the waste systems of over a half dozen cities upstream along the Sacramento River. Much of Orange County's water has already been used in San Bernardino and Riverside Counties, farther up the Santa Ana River watershed.

Orange County has been a world leader in recycling for indirect potable use. In 2015, the Orange County Sanitation District's state-of-the-art plant will begin sending 139 million gallons of such highly treated wastewater to the Orange County Water District, which will put 100 million gallons per day in the coastal groundwater basin. Recharging certain coastal wells with freshwater holds back salty seawater that otherwise moves inland underground and could contaminate the city's drinking water wells. After its ultimate cleansing by natural processes in the time spent below ground, water is ultimately served, once more, to 850,000 residents.

The West Basin Municipal Water District operates the Edward C. Little Water Recycling Facility, the world's largest of its kind. They clean about 30 million gallons of water a day, producing five types of "designer" water from irrigation water to ultrapure water for special industrial needs. The municipal district partners with the West Basin Replenishment District to put treated water below ground to hold back seawater intrusion and for future extraction in domestic water wells. The Replenishment District once relied on imported water for 60 percent of its supplies. Today that portion has been reduced to 20 percent, and they are aiming to rely 100 percent on local groundwater, runoff capture, and recycling in the near future.

Similarly, the cities of Santa Monica and Long Beach intend to stop buying imported water due to recycling and more efficient use. Los Angeles announced recycling efforts in 2008 for treating 4.9 billion gallons of wastewater to drinking standards by 2019. San Diego's "Pure Water Project" plans three treatment plants to produce 83 million gallons per day by 2035 and provide one-third of its supplies. The project could save $1.8 billion by eliminating the need for required upgrades to the old Point Loma Wastewater Treatment Plant. Marin County has 25 miles of underground pipes to deliver treated wastewater to car washes and to toilets in the county jail and in a convalescent home. The return is an average savings of one million gallons a day.

Benefits are increasing for utilities and their customers because treatment costs are now lower than purchasing imported water. San Diego's recycling costs are about 20 percent cheaper, according to the city's website. There are exciting benefits for the statewide environment, too, as pressure is taken off rivers and estuaries that have given up so much, too much, of their life-giving water.

Meanwhile, the biggest percentage of recycled water is still used on farm fields, golf courses, parks, and freeway landscapes. Sonoma County vineyards irrigated with reclaimed water use up to 15,000 AF of recycled water every year. ExxonMobil's refinery in Torrance saves 4,250 gallons every minute by using reclaimed water as a coolant.

Stormwater Capture and Graywater Reuse

With less of California's water predicted to fall as snow, due to climate change, mountain reservoirs may no longer be able to handle all of the rainwater. Western states had a long-standing tradition of making it illegal for individuals to capture and use precipitation. Stormwater falling onto urban areas usually runs rapidly off roofs, impervious parking lots and roads, and into sewers and flood channels. That water can move pollution into rivers, lakes, and the ocean. But since passage of the Rainwater Capture Act of 2012, Californians are encouraged to "slow it, spread it, sink it," by legally capturing and using rainwater. The "new" source of on-site water supply (it has always been there, just underutilized) can reduce the use of precious drinking water on the landscape and recharge underlying groundwater aquifers.

Using cities like sponges by capturing stormwater in the San Francisco Bay area and urban southern California can, conservatively, add more than 400,000 AF a year to the water supply. Hundreds of thousands of AF should also be harvestable in the rest of the state. Rain capturing or "water harvesting" practices can restore the benefits of the natural water cycle to urban areas.

The TreePeople organization and the Council for Watershed Health led efforts in southern California to install rain barrels, rain gardens, underground cistern tanks, and water-permeable walkways and driveways. In Coldwater Canyon Park in Los Angeles, TreePeople built a 216,000-gallon underground cistern that captures stormwater from nearby building roofs and a parking lot. The park uses the water for irrigation during the dry months. In 2010, a three-day storm sent 70,000 gallons of rain into the cistern that would have otherwise flooded the neighborhood. On Elmer Avenue, in Los Angeles, rainy seasons often led to flooding because there were no storm drains. That city block has been retrofitted to maximize the percolation of water back to the ground. Infiltration galleries under the street provide 16 AF of recharge water annually, roughly the amount of water used by 30 residences. Landscaping and native trees appropriate to the climate are part of the project (fig. 100).

TreePeople is also working with the City of Los Angeles to rewrite its stormwater management plan. Next to Sun Valley Park, the city and Los Angeles County plan to convert a gravel pit and concrete plant into a 46-acre park that will collect enough water to supply 4,000 people in an average rainy season.

Individual homes can harvest surprising amounts of rainwater off their roofs into barrels or tanks, which can then be used to water plants. Barrel systems can be inexpensive and easy to install. Some communities provide rebates, as the campaign to increase

Figure 100. Porous surfaces allow rainwater to settle into the ground, instead of rapidly running into gutters, in a yard where drought-tolerant plants replaced thirsty lawns. This was part of TreePeople's Elmer Avenue project in Los Angeles.

rainwater harvesting spreads. For every inch of rain, 1,000 square feet of roof can gather 600 gallons of water. Enter your zip code and roof size of your house in one of many online sites to calculate how much water could theoretically be harvested. The results for my home (in a dry part of California that receives only 11 inches of rain a year) were a surprising 7,880 gallons!

TreePeople's recommendations for homeowner rainwater capture include purchasing barrels or tanks from reputable rainwater tank companies; using barrels with removable lids so they can be cleaned and with a screen that can easily be replaced if torn; avoiding light-colored or see-through barrel materials so light will not cause algae growth; including a metal faucet at the

Figure 101. Rainwater cisterns and barrels can capture hundreds, or even thousands, of gallons of water a year for landscape watering, instead of losing it down storm drains.

base of the barrel, and a metal overflow, and mounting the barrel off the ground to provide gravity flow, strapping it securely in place (fig. 101).

Graywater systems are another area of new possibilities, becoming feasible when changes were made in 2013 to the California Plumbing Code (chapter 16). Graywater comes from washing machines or sink drains and can be used for landscape watering. No permit is required for washing machine systems. Shower systems do need a permit.

Systems *must* have an easy way, such as a three-way valve, to be shut down and the flow directed back to the sewer or septic system; they must send water to irrigate plants on the landscape; must keep the water on the individual's own property; and must discharge the graywater beneath a two-inch cover of mulch, stones, or a plastic shield (no surface pooling or open tubs are allowed).

Systems *must not* contain diaper water or hazardous chemicals, allow people or pets access to the graywater, or connect to the drinking water supply.

The impacts of capturing rainwater in barrels and reusing graywater are big, yet are smaller than from water savings conservation measures like fixing leaks, replacing thirsty lawns with native plants, and simply adjusting watering schedules.

SQUEEZING THE SPONGE: CONSERVATION

Another way to generate "new" supplies of water is simply to make wiser use of water and stretch the existing supply. Conservation measures thoroughly "squeeze the sponge." In the severe drought of 2012 to 2014, the state water board demanded water conservation plans from 400 water agencies. In addition, it required water use reporting for the first time and enabled online comparisons of residential water use in gallons per person per day. In 2015, with the April 1 snow survey finding a record low five percent of normal snowpack in the Sierra Nevada, Governor Jerry Brown issued the first ever mandatory water conservation directive in California history, requiring a 25 percent cut in urban water use.

Conservation measures prove their practicality during droughts, but as soon as droughts end, daily habits tend to revert to extravagance. In conserving, as in recycling, it makes sense to focus less on the extra effort and more on the positive benefits of frugal behaviors and attitudes. Conservation should become a way of life.

The assumption used for many years that 1 AF served the annual needs of five people was actually extravagant. In Los Angeles, where DWP customers have been given modern low-flush toilets for free, an AF today serves eight people. DWP has given away over a million low-flush toilets that use just 1.6 gallons per flush (fig. 102). Older models use 3.5 to seven gallons every time they are flushed. Because toilets account for 40 percent of the water used in households, this retrofitting program allowed the Los Angeles population to grow by 32 percent after 1970, without increasing the amount of water the city consumed. National law now requires toilets that are sold or installed to use no more than 1.6 gallons per flush. If you are still using an old toilet, please consider a replacement. Continuing with an older model, even one that still works just fine, is a great waste of water. Dual flush toilets are widely used in Europe, but just becoming common in the United States. They wisely flush smaller amounts of water for liquids while providing a bigger flush alternative for solids. A recent innovation is the "Stealth Toilet," an ultrahigh efficiency unit using only 0.8 gallons for each very dependable flush.

The Pacific Institute estimates that 400,000 AF of water per year can be conserved by California's urban users by replacing inefficient toilets, showerheads, commercial spray-rinse nozzles,

Figure 102. Publicity for free low-flush toilets for Los Angeles Department of Water and Power customers.

and washing machines. Modern washing machines have been redesigned to improve water and energy efficiency. Front-loading machines do not fully immerse clothes, instead rotating them through a pool of water. They require 25 to 35 percent less water than standard washers, or 16 gallons *less* per load.

Leaky faucets are insidious. They are a pain to repair, but day in and day out, they deliver water straight from the tap to the sewer, without any beneficial use along the way. Sixty drips per minute can waste 190 gallons per month, more than six gallons each day. A "silent leak" in the toilet can keep 30 to 50 gallons a day running through the pipes and down the drain. California water utilities lose 15 percent of their supply to leaks, an estimated 228 billion gallons per year, of 21 gallons per person every day.

Figure 103. A hose running while a car is washed. A spray nozzle can stop this waste of water.

Behaviors and habits can be the most important conservation steps. Running faucets consume two to five gallons every minute! It makes no sense to leave water running while you brush teeth or shave; turn it off after wetting the brush or rinsing the razor, and you can save three or more gallons each time. Running water down the drain while waiting for it to get colder is a thoughtless waste of a precious resource. If you are prone to that mistake, try keeping a cold pitcher in the refrigerator for drinking. Never thaw frozen food under *running* water. Alternatives are to thaw food under water in a bowl or more quickly in a microwave oven.

Water and brooms should not be confused. Five minutes of hosing a sidewalk or driveway can waste 25 gallons of water. Put a shut-off nozzle on the hose for washing the car and turn it off until running water is required (fig. 103). Leaving the hose running for 20 minutes while washing a car can use 100 to 200 gallons of water!

If you wash dishes in the sink by hand, fill the basin, rather than leaving the water running, and save 25 gallons. Do not start dishwashers until they are fully loaded. Familiarize yourself with the dial on the washer and take advantage of the short-cycle option to save three to four gallons per wash. If you use the toilet as a wastebasket or ashtray, consider the better alternatives, but more critically, avoid flushing just *that* stuff. Let it sit until there is a *real* need to flush.

Think about how many showers are taken in your household. Low-flow showerheads cam save 10 gallons of water every time they are used. If someone in your family loves long, slow showers, ask him or her to put a bucket in the shower to trap some of the water, then use the results to start a discussion about how much water is lost down the drain.

In warm parts of California where swimming pools are popular, 100 gallons of heated water may evaporate each day from a single pool, totaling about 3,000 gallons per month. Use a pool cover, not just to help reduce heating costs, but to minimize evaporation when the pool is not being used.

The California Urban Water Conservation Council has identified Best Management Practices (BMPs) and secured a commitment from urban water suppliers to institute BMPs that provide the conservation benefits of low-flow showerheads, low-flush toilets, leak detection programs, metering, tiered pricing, and public information campaigns. BMP implementation aimed at reducing demand by 1.5 MAF by 2020. All new homes built since 1992 in California are required to have meters, which provide the most basic pay-for-what-you-use incentives. Amazingly, a few major cities in California, including Sacramento and Fresno, have still not finished installing water meters for all their customers, because the law set a distant 2025 deadline.

The "Save Our Water" program was created in 2009 by DWR and the Association of California Water Agencies, offering "ideas and inspiration for permanently reducing water use – regardless of whether California is in a drought." The Water Conservation Act of 2009, known as "20×2020," set the statewide baseline for urban water use at 192 gallons per capita per day (gpcd), then calculated a 20 percent reduction target of 154 gpcd. The same act required agricultural water suppliers to adopt agricultural water management plans for measuring water deliveries and pricing structures that encourage conservation and efficiency. Urban retail water and agricultural water suppliers not meeting the water conservation requirements can lose eligibility for state water grants or loans.

Prices need to be set to ensure basic water needs are available to everyone, but tiered pricing can discourage waste by significantly increasing charges as more water gets consumed.

Total potential savings from efficiency estimated by the Pacific Institute jumped as high as 5.2 MAF a year, when water used on landscaping was used more wisely.

A More Logical Landscape

Landscaping appropriate to a dry climate can save fantastic amounts of water. In southern California, 60 percent of water is used for landscaping and half of that is not needed, because most grass is overwatered. A 1,000-square-foot lawn can use 2,100 to 3,600 gallons each month. Lawns should be watered in the evening or the morning, so less evaporates immediately and the water has a chance to soak into the root zone. Depending on your local climate and the season, water every other or every third day. Give yourself or your gardener a break from frequent lawn mowing and set the mower to cut higher; grass grown two to three inches high blankets the soil and reduces evaporation.

The Santa Ana Watershed Project Authority recommends a do-it-yourself approach to save water: walk into the middle of your lawn once a week and pay attention to how the grass feels and sounds under your feet. If it squishes, cut the watering time by half. Keep checking every week and keep cutting back watering time until you feel and hear "crunch." Then add one minute to the watering time. Make adjustments as the seasons change.

Automatic water systems that turn on sprinklers in the middle of rainstorms are a pathetic waste. The Municipal Water District of Orange County has given away thousands of home-sprinkler systems controlled by a sophisticated central system that monitors local weather and automatically sends on–off instructions via pagers (fig. 104).

You could just remove that "crop" that feeds no one—lawn grass—and replace it with vegetation adapted to the local climate. In 2014, 20 million square feet of turf removal rebates were processed, according to the Southern California Water Committee., with cities and water agencies paying $2 to $4 per square foot to homeowners who replaced grass with drought-tolerant landscaping. City and utility programs across Southern California in the Metropolitan Water District of Southern California (MWD) service area are coordinating turf removal through SoCalWaterSmart.com, where step-by-step directions are posted.

Beyond water savings, when lawn grass is replaced with native plants adapted to the state's Mediterranean climate, native birds and insect pollinators will respond, becoming part of a yard's ecosystem, and yard care will require fewer fertilizers and pesticides.

Figure 104. Sprinkling the road, the sidewalk, the driveway, and, incidentally, the lawn. A simple adjustment would save many gallons of water.

Conservation on the Farm

Between 1967 and 2010, applied water use on California crops dropped by almost 20 percent. As the value of crop production rose at the same time, the "economic efficiency" of agricultural water doubled during those years. Yet, because agriculture uses 80 percent of the state's developed water supply, every percentage increase in efficiency translates into tremendous amounts of water. The Pacific Institute calculated that 600,000 AF of water per year could be saved by applying smarter irrigation scheduling on just 30 percent of the state's vegetable and orchard acreage, and converting 20 percent of Central Valley vegetables and 10 percent of orchards and vineyards to drip and sprinklers, and by practicing "regulated deficit irrigation" on 20 percent of current almond and pistachio acreage in the Sacramento Valley. Deficit irrigation applies water during drought-sensitive growing stages, but limits or stops watering during the vegetative stages or late ripening period. By extending efficiency methods more broadly, water savings could be as high as 6 MAF per year, according to studies conducted by CALFED in 2000 and 2006 and by the Pacific Institute in 2009.

The Water Conservation Act of 2009 required agricultural water suppliers serving 25,000 irrigated acres or more to, finally, begin measuring the volume of water delivered to their customers and adopt pricing based, in part, on the quantity of water delivered, as part of Water Management Plans.

SUSTAINABLE GROUNDWATER

Until 2014, California's SWRCB had permitting authority over surface and "subterranean stream water" only. California was the sole state government in the nation without authority over its "percolating groundwater." With land ownership came the right to drill wells and pump without oversight. The extreme drought of 2012 to 2014 and the extreme groundwater overdrafting it engendered, especially in the San Joaquin Valley (map 28), finally opened a door, on September 16, 2014, to passage of the Sustainable Groundwater Management Act. The Act sets a goal of "sustainable yield" in groundwater basins. Boundaries of groundwater basins were defined along with the enforcement powers granted to local Groundwater Sustainability Agencies (GSAs). Though the Act does not give the state power to itself authorize or prohibit groundwater withdrawals, the SWRCB can now intervene if locals fail to act or make inadequate plans (but may not step in before 2025). The seriousness of overdraft conditions

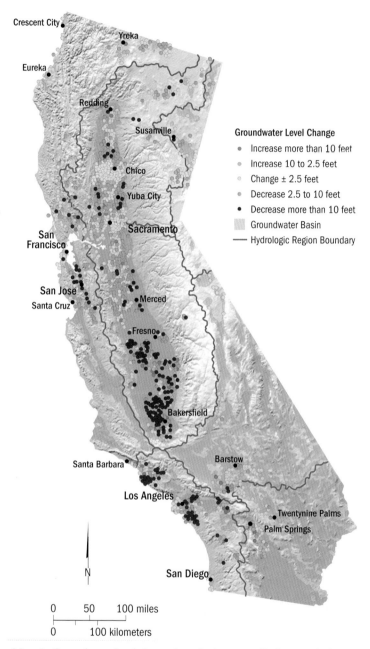

Crescent City

Yreka

Eureka

Redding

Susanville

Chico

Yuba City

San
Francisco

Sacramento

San Jose

Santa Cruz

Merced

Fresno

Bakersfield

Santa Barbara

Barstow

Los Angeles

Twentynine Palms

Palm Springs

San Diego

Groundwater Level Change

- Increase more than 10 feet
- Increase 10 to 2.5 feet
- Change ± 2.5 feet
- Decrease 2.5 to 10 feet
- Decrease more than 10 feet
- Groundwater Basin
— Hydrologic Region Boundary

N

| 0 | 50 | 100 miles |

| 0 | 100 kilometers |

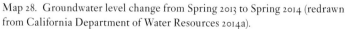

Map 28. Groundwater level change from Spring 2013 to Spring 2014 (redrawn from California Department of Water Resources 2014a).

will determine priority classifications set by the Department of Water Resources. GSAs responsible for high and medium priority basins must submit sustainability plans by January 31, 2020.

This was a historic and controversial change in California's water world. Only one farm organization, representing family farmers, supported passage of the Act. Otherwise farm interests were staunchly opposed, fearing the measure stepped on their property rights and land values. Many other Californians lauded this as long overdue steps toward rational groundwater management.

It is unfortunate that implementation of the new plans was set so far off in time, not until the year 2040.

Water in the Bank: Groundwater Storage

Because we have overdrafted and depleted groundwater basins, an opportunity exists for increased water storage in those aquifers. Groundwater basins can provide "water in the bank" savings accounts to help during dry climate cycles. "Conjunctive use" is the term water agencies adopted to describe the deliberate coordination of surface water with groundwater supplies. Groundwater storage has several advantages over reservoirs. It produces none of the evaporation loss, is ordinarily less vulnerable to pollution, and reduces the impacts of dams and canals on habitat and natural watercourses. Like off-stream reservoirs, water banking does not create more water but adds to the developed supply, stretching its availability and flexibility. Banking can only add "new" water to the state's supply budget if the water is captured during flood events that would otherwise send it out to the ocean, or the groundwater basin is filled using recycled wastewater that would otherwise be sent out to sea.

"What can $2.7 billion buy us?" a research brief by a Stanford think-tank asked in 2014, as the Proposition 1 water bond was coming before voters. Noting that that dollar amount was authorized for "storage," without specifying surface or groundwater, the research concluded that if the funds are spent on groundwater storage, 8.4 MAF of new capacity could be established. That is six times more storage capacity than the 1.4 MAF of new surface storage for the same investment. Constructing dams is much more expensive.

The Kern Water Bank in the southern San Joaquin Valley near Bakersfield became a very profitable groundwater storage facility. It is on 19,000 acres along the Kern River, near Bakersfield, purchased in the 1980s by the SWP. The state never operated a "bank" there but gave the land to the Kern County Water Agency in 1994 (in exchange for forgiveness of payments on 45,000 AF of contracted water). Unfortunately, surplus "contracted water" is part of the "paper water" problem within the SWP, so the state gave up a valuable public asset for water it had not actually been delivering to Kern. The Kern Water Bank has since then been turned over to the Kern Water Bank Authority, a private corporation. The privatized Water Bank's owners have been able to buy cheap "surplus" water to store in the ground and later sell at large profits.

The Water Bank's 30 square miles of valley floor are on an alluvial fan that is ideal for percolating water into the ground; it can absorb over a half foot each day. Theoretically, the bank can store 1 MAF and extract about 240,000 AF in a year. Recovery is handled by 80 wells. To make the systems valuable beyond a local area's needs, interconnections between aqueducts and canals were necessary. The Kern Water Bank can use the Cross Valley Canal and its own canal to send water westward to

Figure 105. Water recovered from storage beneath the ground at the Kern Water Bank, to be sent to the California Aqueduct for customers farther south.

the California Aqueduct or eastward toward the Friant-Kern Canal (fig. 105). One benefit of the program has been reestablishment of intermittent wetlands habitat in that part of the San Joaquin Valley, which had been converted to farmland. Native upland habitat is also being reestablished on one-third of the bank's lands.

But the Kern Water Bank has remained controversial ever since the state "giveaway" two decades ago, as many critics see that history. There has been litigation. In 2014, a California court struck down the 2010 environmental review done by the state for the water bank, holding that regulators had not examined how its operation affects the state's water resources and wildlife. DWR must conduct another review.

The MWD is a client of the Kern Water Bank and of the SemiTropic Water Bank, also in the southern San Joaquin

Valley, with 350,000 AF of potential storage in Kern County. Other conjunctive-use projects in groundwater basins scattered around the southern state give it another 437,000 AF of additional potential storage capacity. A controversial MWD groundwater-banking project was proposed in the Mojave Desert in eastern San Bernardino County. Using private land owned by the Cadiz Corporation, the MWD would have paid to store underground up to 1 MAF from the Colorado River, but also planned to pump "native" water from that groundwater basin. This was one of the ways the MWD hoped to adapt to the federally mandated 4.4 Plan, as California tried to live within its allocation from the Colorado River. Concerns developed over the Cadiz Valley Water Conservation, Recovery and Storage Project because dry-year extraction of up to 150,000 AF might have exceeded the natural recharge capacity and dry up springs in the Mojave National Preserve. U.S. Geological Survey scientists estimated that the pumping rate proposed might be 25 times the rate of replacement by precipitation. The proponents pledged to monitor the situation with wells, but skeptics worried that once the pumps, canals, and wells were in place, thirsty desert creatures in that distant valley would not have as much pull as millions of thirsty Southern Californians. The situation had disturbing parallels to Los Angeles's export of Owens Valley surface water and groundwater. In 2002, the MWD board voted to cancel this project, but the idea did not go away and further environmental documentation was prepared in 2011. Monitoring and management of groundwater resources were promised to ensure that minimal significant impacts occur.

In 2014, a court rejected lawsuits with challenges to the Cadiz EIR/EIS. In November 2014, the Santa Margarita Water District established the Fenner Valley Water Authority to control delivery of the Cadiz groundwater. Santa Margarita wanted to reduce its reliance on MWD by purchasing water from the project. Cadiz is trying to reach an agreement with other southern California water agencies interested in buying water from the project.

The San Francisco Public Utilities Commission (SFPUC), the California Water Service Company, and the cities of Daly City and San Bruno announced, in December 2014, a management plan for the South Westside Groundwater Basin on the San Francisco Peninsula. The two utilities and two cities agreed not to draw from the basin during wet years, allowing it to replenish and be available during droughts. The SFPUC will supplement the supplies of the other water retailers with free Hetch Hetchy water during wet years.

THE DEBATE OVER DAMS

The $2.7 billion designated for storage in the Proposition 1 bond act passed in 2014 may be spent on either surface storage behind dams or in storage underground. The California Water Commission will decide where the bond funding will be spent as competing projects are submitted. No more than 50 percent of a project's costs can come from the bond, so interested parties will have to provide matching funds.

Dams are controversial. Senator Dianne Feinstein, in an opinion piece in the *San Francisco Chronicle* (August 1, 2001) wrote: "There are those who would challenge any effort to increase the amount of water stored during wet years for use during dry years as potentially damaging to the ecosystem. And there are those who want iron-clad assurance a full allocation of water will be available for agriculture and urban use no matter what

Figure 106. Red Bluff Dam with the gates up to allow salmon to pass.

damage this could cause to the environment. We can provide more off-stream water storage and restore the endangered ecosystem. It should not be one or the other."

Some of the problems that dams create for salmon have been addressed through modifications to existing structures and changes in the way dams are operated (fig. 106). At Shasta Dam, a temperature control device allows operators to pull water from different depths, rather than solely off the warm surface of the reservoir. Salmon require cold water, around 56 degrees F, during critical times when eggs hatch and fry emerge. To ensure that cold water can be released from April through September, more storage must be held in Shasta Reservoir.

Between 1995 and 2010, 51 hydropower licenses affecting 212 dams came up for their first renewal. Hydropower licenses were issued by the Federal Energy Regulatory Commission (FERC) for periods of 30 to 50 years. Though hydropower is a clean form of energy production, diversions through power plants impact streams. Some have diverted 90 percent of the natural flow in rivers. Relicensing presented an opportunity to adjust operating requirements to ecologically based in-stream flow needs. The trade-off is reduced power generation capacity. The process requires studies on fish, water temperatures, passage of sediments and gravels, and historic flow patterns and establishes minimum in-stream flows that mimic nature and are supported by science. CalTrout called the FERC relicensing process "the single greatest opportunity to restore coldwater fisheries in California over the next decade."

Build More Behemoths?

The era of major dam building has ended; all the feasible and available sites for large dams in California's river canyons have been developed. The largest dams were completed between the 1930s and 1970s, but even since 1979, over 1,600,000 AF of storage has been added, including New Spicer Meadow, Warm Springs, Diamond Valley, and Los Vaqueros Dams.

New storage upstream from Friant Dam in the San Joaquin River watershed is being considered. Temperance Flat Dam would be constructed on a site that was considered when Friant Dam was first proposed, seven miles upstream of Friant and partially within the upper end of its reservoir. Though it could hold 1.26 MAF in storage, the actual average yield would be only a modest 180,000 AF per year, at best. The difference between "capacity" and "yield" is the difference between what a reservoir can hold when full and how much water a reservoir can "produce," or yield annually. The sustainable yield that can be taken out of a reservoir, over time, is no more than what the river can annually deliver. Dams do not make more water, but simply allow flexibility about when water is used. In this case, Millerton Reservoir already exists behind Friant Dam on the same river, with its 520,000 AF capacity.

The Bureau of Reclamation's "Upper San Joaquin River Basin Storage Investigation" draft feasibility report for the Temperance Flat Dam was released in January 2014. The new storage might figure into restoration efforts on the lower San Joaquin River, along with management of agricultural water downstream, but there are skeptics about the economic benefits forecast for the Temperance Flat reservoir, which may explain why no group or water district has stepped up to cover its nonpublic construction costs.

The last major federal CVP Dam was the New Melones Dam on the Stanislaus River, completed in 1979. One federal project, the Auburn Dam, was halted 25 years ago but became the "neverending story" in water politics. The dam was originally proposed to store water, generate electricity, and create a recreational reservoir on the north and middle forks of the American River. The first design, for a 700-foot-tall thin-arch dam, was abandoned because of earthquake risks. Later plans called for a smaller, redesigned dam solely to reduce the risk of floods reaching Sacramento. It would have cost over $1 billion. Friends of the American River opposed that proposal, pointing to more than 500,000 recreational users of the river canyon and its whitewater rafting stretches every year. They also claimed there was a greater likelihood of a reservoir-induced quake than of the 500-year flood the dam was supposed to prevent.

The Auburn Dam debates generated more of the hyperbole that has characterized California's long history of water development. On December 31, 2001, the *Sacramento Bee* quoted a former Nevada County supervisor as saying, "This state is not going to survive unless we impound more water. We need Auburn dam." Despite such dire warnings, the Bureau of Reclamation, in fall 2002, did something historically rare for that agency: it closed the tunnel that diverted water from its dam construction site, 35 years after initial construction began in 1967, and let the river run again through its historic channel. In 2008, news came that the "never-ending story" actually had reached an end. The SWRCB canceled the unused water rights for that Bureau of Reclamation project. Some saw this as a signal of a major change of values and attitudes about rivers in California.

Off-Stream Dams?

With the best dam sites along rivers long since developed, off-stream storage reservoirs, filled by pumping water from water sources into dry canyons, have been constructed to increase water storage. Off-stream dams provide management flexibility, making water available at different times and places than nature provides it. San Luis Reservoir, used jointly by the SWP and CVP to store water south of the Delta, is one example.

In 1999, the MWD completed the huge off-stream Diamond Valley Reservoir south of the San Bernardino Mountains, near Hemet, with a 12-mile tunnel through the mountains to deliver water to the coastal plain. Diamond Valley can hold a six-month supply for the MWD's customers, should earthquake, terrorism, or drought interrupt flow along the aqueducts (fig. 107).

The DWR and CALFED studied the possibilities for a dam near the town of Sites, in Colusa County, to store up to 1.4 MAF. This "North-of-the-Delta Offstream Storage" (NODOS) reservoir would be filled from the Sacramento River during wet seasons for use during dry periods, by tying into existing irrigation

Figure 107. MWD's Diamond Valley Reservoir, near Hemet. This reservoir was filled in 2002 with water from the north (the SWP) and from the east (the Colorado River).

district canals for supply and releasing back to the river via a new pipeline. Construction costs could be as much as $4.1 billion, so the water would be at the expensive end of prices paid in Southern California. Rather than cheap water, the dam is intended to provide flexibility in water management with environmental benefits possible for Delta water quality and fisheries protection flows. A concern about the project is that reducing high runoff in the Sacramento River may further reduce natural cleansing and gravel movements important to migratory fish. The limitations to "surplus" water for supplying off-stream dams become a real concern, considering that drier conditions are forecast for the state due to climate change. For example, during the severe drought years of 2012 to 2014, the San Luis reservoir fell to less than 17 percent of its capacity, at one point, because less water was available for that off-stream reservoir.

Raise Existing Dams?

An off-stream reservoir east of San Francisco Bay, called Los Vaqueros, was enlarged by the Contra Costa Water District in 2012. By raising the dam's height 34 feet, reservoir capacity expanded from 100,000 to 160,000 AF. There is discussion about pursuing a further raise of this dam.

For more than a decade, the U.S. Bureau of Reclamation has been considering possibilities for enlarging Shasta Dam. Three enlargement options were evaluated in a draft feasibility report in 2011: raising the crest of the dam by 6.5, 12.5, or 18.5 feet. Those height increases would increase reservoir storage by 256,000, 443,000, or 634,000 AF, respectively, with construction costs ranging from $891 million up to $1.15 billion. Besides the expected increase in water supply for downstream users, the options promised improvements for Sacramento River salmon, by adding colder water downstream to improve spawning conditions. An 18.5 foot raise of the dam would, however, inundate another half mile of the lower McCloud River, which lost 15 of its 35 miles when the original reservoir filled.

The McCloud River is an area of specific interest, because Proposition 1 (2014) prohibited funding projects that "could have an adverse effect on the values upon which a wild and scenic river or any other river is afforded protections pursuant to the California Wild and Scenic Rivers Act or the federal Wild and Scenic Rivers Act." With sacred sites and burial grounds of the Winnemem Wintu Indians already covered by the reservoir and additional sacred sites upstream, further impacting the already over-allocated Sacramento River is a problem.

The city of Sacramento will have better flood protection along the American River when the Army Corps of Engineers

Figure 108. Sacramento River, confined between levees to protect homes and farms from floods.

finishes raising Folsom Dam by 3.5 feet, and extreme flood events will be handled more safely by widening the emergency spillway gates. Sacramento had already substantially increased its degree of flood protection after 1986 storms came close to overwhelming levees. Nineteen miles of the American River levee system were improved, increasing protection to a 100-year flood level (fig. 108).

Raze Existing Dams?

Dams are not forever. Each has a life expectancy; sediments will fill them, in time. Given the impacts dams have had on some species in California, it is important to keep in mind that "extinction is forever, dams are not." Whenever a dam is considered for demolition, cleanup and handling of the sediment trapped behind it must be addressed. The costs of removal have to be balanced with the costs of attempted repairs or improvements to dams nearing the end of their useful lives. For very large dams, renovation can be very difficult.

Between 1920 and 1956, 22 dams were removed along 100 river-miles in the watershed of the Klamath River, a North Coast river that carries heavy sediment loads. In 1969, the city of Eureka blew up Sweasey Dam on the Mad River, which had completely filled with silt.

The fate of four old Klamath River dams (in Oregon, upstream from California), which impede salmon runs, was still unresolved in 2015, after a 2014 deadline passed for Congressional approval of settlement agreements between farmers, tribes, PacifiCorps (the operator of the hydroelectric generation plants), and the government. The plans called for decommissioning and removing the Iron Gate, J.C. Boyle, Copco 1, and Copco 2 Dams by the year 2020, taking all four out in one year. That will open up 300 miles of salmon spawning and rearing habitat. A California Energy Commission report had found that it would be cheaper to remove the silt-loaded dams than build the fish ladders needed to reestablish blocked salmon runs.

Small irrigation diversion dams were removed in 1998 from Butte Creek, in the Sacramento Valley, to restore 25 miles of unimpeded flow for migrating salmon. Farmers kept receiving irrigation water through redesigned systems, but the creek itself again flowed free. During the following migration seasons, thousands of fish used the opened channel. Secretary of the Interior Bruce Babbitt helped celebrate the 1998 dam removal (fig. 109). He told reporters, "We're not taking aim at all dams, but we should strike a balance between the needs of the river and the demands of river users... In all probability the process will continue on a dam-by-dam basis, with states and community stakeholders making most decisions. But there can be no

Figure 109. Secretary of the Interior Bruce Babbitt using a sledgehammer to begin the historic dismantling of a small dam on Butte Creek in the Sacramento Valley in 1998.

doubt that we have a long way to go toward a better balance" (Friends of the River 1999, 2).

The largest dam removal project in California moving forward in 2015 was on the Carmel River, where the reservoir behind the San Clemente Dam had become 95 percent full of mud, was declared a seismic hazard, and, in 2002, taken out of commission. The sludge backed up by the dam loomed upstream above 1,500 homes. Instead of hauling away 2.5 million yards of sediment, the river has been diverted into a new half-mile-long channel. After the dam is taken down, the sediment field will be revegetated. Steelhead Trout will again have access to eight miles of spawning habitat that had been blocked since 1883 (steep fish ladders were never effective at this dam).

For over a century, Stanford University's antiquated Searsville Dam had an enormous negative impact on San Francisquito Creek watershed and the greater San Francisco Bay estuary. Built

Figure 110. "Restore Hetch Hetchy" bumper sticker.

between 1890 and 1892, the 65-foot-tall dam lost over 90 percent of its original water storage capacity as 1.5 million cubic yards of sediment filled in the reservoir. Searsville Dam does not, today, provide potable water, flood control, or hydropower benefits. With the dam removed, the valley can again provide flood protection and steelhead spawning habitat on the creek's largest tributary, flowing through Portola Valley and Woodside.

In 2000, 87 years after Congress authorized construction of a dam in Hetch Hetchy Valley, the head of the Restore Hetch Hetchy organization authored "The Hetchysburg Address," which opened, "Four score and seven years ago, our fathers brought forth upon this continent a new dam and reservoir, conceived in the Bay Area, and dedicated to the proposition that all national parks are not created equal. Now we are engaged in a great debate – testing whether that dam and reservoir, or any things so conceived and so dedicated – should any longer endure" (fig. 110).

Restore Hetch Hetchy? What would San Francisco and the other Bay Area cities using Tuolumne River water do for a supply if O'Shaughnessy Dam were taken down? As shocking as the idea seemed, in 1988 Secretary of the Interior Donald Hodel suggested that the restoration of Hetch Hetchy might be feasible *without* reducing the city's water supply. The concept is not to take the SFPUC's water away, but to replumb the Tuolumne River water delivery system, taking advantage of storage opportunities in New Don Pedro Reservoir, downstream from Hetch Hetchy, the Calaveras Reservoir in Alameda County, and the Los Vaqueros Reservoir in Contra Costa County.

Restoration of the scenic valley in Yosemite National Park would require removal of relatively small amounts of sediments behind the dam, as the Tuolumne River basin transports so little material off that granite-dominated watershed. The report prepared for Interior Secretary Hodel concluded that "within two years extensive areas on the floor of Hetch Hetchy Valley would be covered with grasses, sedges, rushes, and other herbaceous plants that would be highly visible." It would take longer for trees and shrubs to grow, and it might take decades for the "bathtub ring" to disappear along the canyon walls. Full restoration could take a century, but it could be accelerated by active restoration measures. The Hodel report noted that there would be less electrical output from hydroelectric plants in the SFPUC system, but the first priority of the Hetch Hetchy system has been water supply. In droughts, water has often been held back though it meant cuts in electricity generation.

The Restore Hetch Hetchy organization hoped to have dam removal under way by 2013, the 100th anniversary of the Raker Act, but their quest must continue. Restoration of Hetch Hetchy Valley to Yosemite National Park would be a gift to the nation and to the world, an amazing gift to our children and grandchildren. John Muir would be enormously pleased.

TRANSFERS: WATER AS A COMMODITY

Every year hundreds of water transfers occur in the state. Water marketing accounts for roughly five percent of all water used annually by California's businesses and residents (about 2 MAF of water trades are committed annually, with around 1.4 MAF in actual flows exchanging hands). Over time, the market has shifted from primarily short-term (single-year) contracts to one dominated by longer-term and permanent trades. Farmers are the primary source of water, and the destinations include other farmers, cities, and the environment. Market growth has slowed in the last decade, reflecting a variety of constraints, including new pumping restrictions in the Sacramento–San Joaquin Delta and more complicated approval procedures. In 2008, dry conditions were behind purchase of 230,000 AF from northern California farmers by buyers south of the Delta. The total quantity made available for transfer in 2012 was 188,074 AF and in 2013, 268,370 AF.

In 1991, in the fourth year of an extended drought, Governor Pete Wilson issued an executive order creating the California Drought Water Bank. The bank was an emergency drought measure to coordinate the sale of surface water to needy buyers. The bank purchased surface water from willing sellers and sold it to buyers experiencing critical shortages. The water was made available by fallowing farmland and substituting groundwater for the marketed surface water. The bank operated in 1991, 1992, and 1994. It was not a permanent solution to the long-term reality of droughts, and definitely not something that communities or agencies experiencing shortages could count on when planning development.

In 2005, the MWD negotiated an agreement to pay Palo Verde Valley farmers (along the Colorado River near Blythe) to stop growing food, releasing up to 118,000 AF of water each year to the MWD. Palo Verde stops irrigation on up to 28 percent of their land in a given year, based on MWD requests. Over the 35 years of the plan, the district may spend up to $337 million to secure 3.6 MAF of water. The deal did not provide any "new" state water for the MWD but replaced some of the "surplus" Colorado River water that California must stop using. The MWD negotiated individual deals with farmers who held the water rights. They made one-time payments of $3,170 for each acre set aside. Agriculture has been the primary economy in the valley, where 37,000 people resided. The MWD said that the impact on the local farm economy would be minimal, because only the least productive land would be set aside. Farmers were not so certain, but went ahead with the deal, in part because they perceived the political clout of urban areas to be so great that their water might be taken away forcibly, if the societal decision was that urban needs were the "highest use."

Other farmers had learned a lesson about urban clout when they resisted pressure to transfer water to San Diego. When the December 31, 2002, deadline arrived without an agreement on the 4.4 Plan to wean California from its overuse of Colorado River water, the Department of the Interior announced immediate cuts for the MWD (instead of the 15-year gradual cutback that a plan would have triggered). The department also, however, declared that the Imperial Irrigation District (IID) must shift 200,000 AF of its historic water allotments to the MWD. The IID had previously balked at a plan to annually sell up to 200,000 AF to the San

Diego County Water Authority. The IID board had objected to the land fallowing that this sale required. Despite proposed payments to farmers, cuts in acreage would echo through the valley's economy. The IID board also wanted water interests pushing the transfer to share some of the district's liability for the damages that would result to the Salton Sea environment when fallowing reduced the amount of water reaching the sea.

The competing pressures of agriculture, urban growth, and the environment had never been so clear. One plan for Salton Sea "restoration" suggested that lesser amounts of water flowing to the sea be concentrated into a smaller area, confined by dikes, where salinity suitable for fish and waterfowl could be maintained. The main portion of the sea would become hypersaline, something akin to Mono Lake.

Historically, Californians have not paid for water itself. Paying a water bill feels like buying water, but the bill actually covers only the costs of storage, treatment, and delivery to the tap. Water in California cannot be "bought" because all of it is owned by the state. There have long been legal obstacles to moving water from agriculture to urban users. The state legislature and the federal Central Valley Project Improvement Act (CVPIA) cleared the way, so that water can be marketed as a commodity. State Water Code provisions still require that there be no adverse effects on other water rights holders and no unreasonable effects on fish and wildlife resources.

Given that water is essential for life, it is amazing how little philosophical debate preceded this move into the market. The philosophy that the free market can best manage anything is being given a test. Several philosophical questions deserved more attention than they received before the obstacles were removed:

- If water is to be sold only to those who will meet a seller's price, where does that leave the poor? And how will the environment compete?

- How can we balance the vital life-giving functions of water in the California landscape against the use of water simply for convenience or comfort?

- How much care will be taken to assure that real water is being marketed, rather than "paper water," which may be based on water rights that cannot be satisfied?

- Do we want to encourage "water ranching" by land speculators who acquire real estate just to sell water taken away from it?

Because water is vital for life, governments must manage the resource for the common good. Regulation is needed to protect the public trust. It remains to be seen if the California Human Right to Water Act of 2012 will take the state in a new direction. "In California," a saying goes, "water flows uphill toward money." Norris Hundley Jr.'s pessimistic conclusion (in *The Great Thirst*, 2001, 519) was: "With water, like land, now subject to a market system and allowed to go blindly to the highest and most powerful bidders – invariably metropolitan areas and developers – the future holds little promise of being fundamentally different from the past."

CLEAN WATER

The Human Right to Water Act of 2012 promises that every person will have "safe, clean, affordable, and accessible water adequate for human consumption, cooking, and sanitary purposes."

We all can help minimize the pollution generated by California's large population by picking up after our pets and disposing of droppings either in the toilet or in trash that goes to a landfill. We should keep outside trash cans covered, keep leaf piles away from gutters and drains, minimize or stop using pesticides on lawns and gardens, never hose down sidewalks or driveways, and support community projects to divert draining water into the ground, the route it once naturally followed.

In 2002, the Los Angeles RWQCB adopted new requirements for water quality in Santa Monica Bay. It was finally complying with a 1979 deadline set in the federal Clean Water Act, in response to legal pressure from organizations such as Santa Monica BayKeeper, the NRDC, and Heal the Bay. The requirements allowed another three years for meeting state bacterial standards at beaches in the summer, and six years to achieve winter dry weather standards equivalent to natural conditions. The standards apply to all beaches in Ventura and Los Angeles Counties.

Achieving cleanup of storm runoff has been a much tougher accomplishment. The IRWD is building artificial wetlands to clean up contaminated water. The success of such efforts was demonstrated in Humboldt Bay, far to the north. Wetlands work as well as traditional drainage and treatment facilities and cost less in the long run. They also can provide side benefits as wildlife habitat and bird-watching spots for humans. The new emphasis on rainwater capture, across cities and on each home site, can go far to solve this problem.

The Chino Basin dairy water pollution issues had been addressed by trucking manure out of the basin, but the Inland Empire Utilities Agency also built a desalination facility for processing 12 million gallons per day to remove the nitrates in groundwater saturated by cow wastes. Manure itself helps power the desalination plant to start an elegant cleanup and recycling chain: methane gas is harvested from composting cow manure and used to generate electricity, which runs the desalter, which augments the local water supply.

In 2014, the Irrigated Lands Regulatory Program modified the agricultural waivers that had been granted decades ago when the Clean Water Act was established, by now requiring monitoring of runoff pollution, meeting tougher standards, and on-farm inspections by regulators.

ECOSYSTEM RESTORATION

Faced with the legal precedents established at Mono Lake, in December 2006, a historic event occurred in the long saga of contention between Los Angeles and the Owens Valley. Sixty-two miles of the lower Owens River, dry since the city had diverted the water into the Los Angeles Aqueduct a century earlier, were rewatered. A court had earlier begun fining the Los Angeles Department of Water and Power $5,000 a day for delaying implementation of a river rewatering plan. The stretch of the Owens River is coming alive again with riparian vegetation and wildlife. River water is pumped back into the aqueduct just before it reaches the dry Owens Lake bed.

As a result of separate legal requirements to address the air-polluting dust storms off the salty, dry lakebed, Los Angeles Department of Water and Power began spreading water with a system of delivery pipes and "bubblers" in 2002 (fig. 111). About 25 billion gallons of water were annually applied to 22.5 square miles of the Owens Lake bed. The city spent $1.3 billion after the year 2000 on controlling dust, but had still not met a 2006 deadline to show compliance with clean air standards. In November 2014, the city and the Great Basin Unified Air Pollution Control District settled on an agreement using waterless

Figure III. Dust control bubblers putting water onto the Owens Lake bed to control dust storms that violate air quality standards.

dust control: they will plow three-foot high furrows on 50 square miles of the lakebed which, experiments show, will adequately reduce windblown dust.

Legal precedents set by the Mono Lake decisions also applied to restoration of salmon runs and natural river functions on the lower San Joaquin River. The Friant Water Users Authority and a consortium of 16 environmental groups, led by the NRDC, agreed to work together toward a settlement plan to meet restoration goals without costing the water users any money or water (fig. 112). The Vernalis Adaptive Management Plan has experimented with 31-day pulse flows in the spring, to benefit juvenile salmon migrating out to sea, and October "attraction" flows to help returning spawners locate their home waters. (Vernalis is the location where the San Joaquin River, with all of its tributary waters, enters the Delta.)

Figure 112. The San Joaquin River, one of California's largest river arteries, bone dry where it used to flow beneath Highway 152 near Los Banos. The river was dewatered below Friant Dam to serve agricultural irrigators, but is being restored.

Efforts to restore 100,000 acres of San Francisco Bay's tidal marshlands were aided in 2002 when commercial salt ponds in the South Bay were purchased by the state to restore waterfowl and wildlife habitat (fig. 113). Progress was complicated because mercury had washed down from mines upstream and settled into the bottom mud. Restoration of more salt ponds in the Napa marshes, at the north edge of the bay, began in 2008. Levees around the series of ponds, now managed by the Coastal Conservancy, were breached. Salt production, by evaporation, had been going on since 1956 on wetlands that were once rich with migratory birds and tidal life. Old aerial photos and maps were used to locate historic channels to determine where to bulldoze levee openings. Within two years, the marsh ecosystem was recovering and again attracting migratory waterfowl.

Figure 113. The southern end of San Francisco Bay. Red and green mark commercial salt ponds. Boundary lines roughly delineate more than 16,000 acres acquired in 2002 to be restored as wetlands habitat.

In Humboldt Bay, in Northern California, water quality degradation had nearly destroyed natural values, but today 94 acres of wetlands are maintained by an innovative use of partially treated sewer effluent. Marsh vegetation thrives there as it cleans the discharged sewage water.

Restoration efforts have "captured the hearts and minds of citizens throughout California," according to the SWRCB publication *Opportunity, Responsibility, Accountability: Nonpoint Source Pollution Control Program* (2001, 41). Understaffed state and local agencies increasingly rely on citizen volunteers to help with water quality monitoring. To document the effects of sediments entering Humboldt Bay from logging in that watershed, local volunteers, members of a group called "Salmon Forever," annually collect more than a thousand water quality samples from creeks and rivers entering the bay.

The "Storm Water Detectives," a group of Lodi high school and middle school students given professional training, monitor insects and worms in the Mokelumne River. Their bioassessment data serve as one measure of river health. Other volunteers survey aquatic insects in the Truckee River downstream from Lake Tahoe and bacteria levels in the Yuba River. "DeltaKeeper" members identified very high levels of bacteria in popular swimming and water-skiing areas in the Delta, which led to federal listing of those waters as "impaired" (a Clean Water Act designation that triggers pollution management measures).

Such volunteers, according to the SWRCB, are "learning that streams are inseparable from their riparian corridors, and riparian corridors from their uplands – the watershed. All evolved together – each influenced by and influencing the other. They have come because they realize that whatever happens on the land eventually shows up in the stream at their feet, and they cannot heal the stream without healing the land" (2001, 41).

LEMONADE FROM LEMONS: IS DESALINATION VIABLE?

Can desalination plants augment the entire state's water supply by tapping into the ocean water off the California coast? That idea is raised in every discussion about water limits, it seems.

Most desalination plants in California are similar to the Inland Empire Utilities Agency's: they are inland and do not process seawater. Instead, they treat brackish groundwater with high levels of dissolved solids from sewer effluent or other contaminants. These systems require less expensive filter membranes than do plants that treat seawater (map 29).

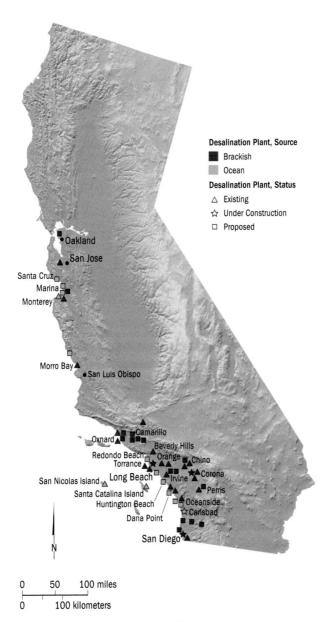

Map 29. California desalination facilities. Most are inland using
brackish water. Only three ocean plants now exist; the Carlsbad
plant will be completed in 2016 (redrawn from California Department of Water Resources 2013a).

Avalon, on Catalina Island, was the first California town to desalinate seawater. The city of Santa Barbara built a desalination facility in 1992 capable of producing 7,500 AF per year. They decided to invest in the plant at the end of a long drought, but ran it for only three months. It has been shut down ever since, because cheaper water became readily available when the drought ended. Some of the equipment was subsequently sold, leaving a current capacity of 3,125 AF. After three years of extreme drought, as a last resort, Santa Barbara was considering bringing the plant back in 2015. The estimated cost was $40 million to restore operating readiness.

Desalinated water typically costs about $2,000 per AF. The cost is about double that of water obtained from building a new reservoir or recycling wastewater, according to a 2013 DWR study. Its price tag is at least four times the cost of obtaining "new water" from conservation methods, despite costs for desalination dropping as new reverse-osmosis membranes have been developed. The MWD gives its member agencies subsidies of $250 per AF to help the developing technology compete with wholesale prices for imported water.

Desalination has substantial impacts that many people have not recognized. The process requires lots of energy; each AF produced requires from 2,500 to 29,500 kilowatt-hours of electricity. It takes about two gallons of seawater to produce each gallon of freshwater. Along with freshwater, a concentrated brine waste is generated (fig. 114). Testing continues on ways to reduce daunting impacts to ocean life from entrainment and from the return of concentrated brine wastes to the sea.

Note that the potential contribution from desalination forecast in the State Water Plan was quite small when compared to opportunities for water supplied by recycling, better groundwater

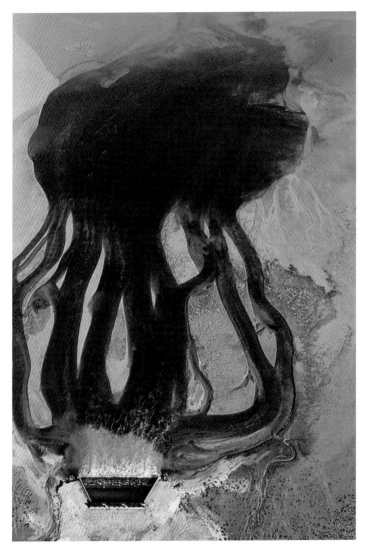

Figure 114. Waste from a desalination plant in Kuwait. Seawater desalination produces 75 percent of Kuwait's water supply. After treatment, concentrated brine wastewater is returned to the sea, where it mixes into the Persian Gulf, creating the shape of a tentacled monster (from Yann Arthus-Bertrand et al., *Earth from Above*).

management, and urban water-use efficiencies (see fig. 98). Santa Cruz city officials shelved plans for a desalination plant because of negative community reaction about growth fueled by more water supply. Marin County studied a desalination project, then dropped it when water use declined. Long-running plans to build a plant in San Francisco Bay near Concord were ended when the region's largest water districts decided they could obtain water more cheaply through recycling and other means. Australia spent more than $10 billion building six huge seawater desalination plants during a severe drought from 1997 to 2009. But as of 2015, four remained shut down because the cost of the water became noncompetitive.

In 2006, there were 21 desalination proposals along the California coast. Since then, only two new projects, in Sand City and in Cambria, have been built. Since 2002, West Basin Municipal Water District has been desalting ocean water at a pilot facility in El Segundo, used for research and water quality testing. West Basin is planning a temporary ocean-water desalination demonstration facility in Redondo Beach to research ocean-water withdrawal and return methods, including using the ocean floor to naturally filter the water as it is taken from the ocean and using screens to minimize impacts on the offshore environment. West Basin's goal is 20 million gallons a day of desalinated drinking water by 2020.

Oceanside built a desalting plant in 1994, prompted by the drought of 1987 to 1992 and by MWD price increases. Its Mission Basin Groundwater Purification Plant does not directly desalt the ocean but draws groundwater affected by seawater intrusion to produce 2.2 million gallons per day (seven percent of the city's needs). Oceanside sends its brine waste 1.6 miles offshore.

Cambria, on the central coast, brought a desalination plant online late in 2014 that does not process ocean water, but a combination of brackish estuary water and recycled wastewater.

Farther down the coast, the city of Carlsbad and the San Diego County Water Authority are going to desalinate ocean water at the Cabrillo Power Plant in Carlsbad. The plant is coupled with an existing power plant that already has seawater intakes for cooling. Poseidon was required to build 66 acres of wetlands in San Diego Bay to offset the plant's environmental harm. It also must blend its brine at a 5:1 ratio with other seawater before flushing it back into the ocean to reduce harm to marine life. The plant will use an enormous amount of energy—about 38 megawatts, enough to power 28,500 homes—to force 100 million gallons of seawater a day through filters. In a 30-year agreement, the San Diego County Water Authority will purchase the entire output of the Poseidon plant, about 9.3 percent of the county's domestic needs. Construction on the plant and pipeline is to be complete in 2016.

Plans by Poseidon to build another desalination plant in Huntington Beach slowed when the Coastal Commission said it wanted the company to investigate whether its pipes could be buried, a prospect that will increase costs. In 2015, the Orange County Water District, which sells water to local retail agencies, said it was interested in buying all of the 56,000 AF of drinking water that may be produced by a Huntington Beach plant. Like groundwater, seawater can be a local supply source that does not require rivers to be dammed and long-range aqueducts to be built. Tapping into the "endless" ocean feeds dreams of making water limits irrelevant. If we succeed, the question becomes, do we really want the unlimited growth that could be accommodated with "unlimited" water?

WILL THERE BE ENOUGH WATER?

Integrated Water Management

Every five years the California Department of Water Resources updates the California Water Plan. The first water plan focused on how to fully develop the state's water resources. Today the focus has shifted toward sustainable water management. The difference has been characterized as "concrete versus the soft path." One new emphasis is increased reliance on local water supplies, rather than on long-distance transfers. The logic of focusing on dry years when estimating available water supply has penetrated. Planning urban development or agricultural crop choices based on water available in "average" years will lead to certain shortages, given the reality of droughts in California's historic climate pattern. Climate change was addressed for the first time in 2003 and has become an increasingly important component ever since.

Water resources often span regions and jurisdiction boundaries. The overarching theme of California Water Plan 2013 became "Integrated Water Management" (IWM), aiming for agencies, organizations, and stakeholders to improve efficiency and productivity through collaboration. Constituencies that have not always been included in water planning, such as Indian tribes, are being drawn into the processes. Goals call for federal, state, regional, and local agencies to seek common understanding and improve efficiency in land use planning, environmental restoration, and groundwater management. Projects that have broad benefits may be able to use funding sources outside traditional water bonds and water agency budgets. Daily progress toward these laudable, if complex, IWM objectives occurs at four DWR regional offices that provide technical guidance and assistance to local and regional organizations. Regional Offices also collect and analyze regional water resources data in support of many DWR programs.

What Future Do You Choose?

Even before the effects of the changing climate became so apparent, supply and demand were out of balance in California. The state is today dealing with "peak water," feeling the impact from actually exceeding supply limits in our cities, on farms, and in the environment. Despite that crisis, there is a lot of water in California. There is enough to serve human populations far greater than today's; we could grow past the 50 million forecast for mid-century to, theoretically, well over 200 million people! That extreme scenario would require taking every drop of water currently serving the environment and farms. Whether we want to consciously, or simply through inertia, keep heading toward such a future and whether anyone would want to live in such a society are important questions. Consideration of such an outlandish water policy can clarify our thinking about more realistic futures.

When California became the most populous state in the nation in 1962, Governor Edmund G. Brown called for a three-day "California First Days" celebration. After hearing about Brown's proclamation, former governor Earl Warren said, in a speech the next day, that he thought the governor was mistaken in his assessment of the importance of mere growth, that there was no merit in simply being the largest. Warren later recalled how he "told them that instead of dancing in the streets, we should...call the people of California to the schools, churches, city halls, and other places of public assemblage, there to pray for the vision and the guidance to make California the finest state in the Union as well as the largest" (Warren 1977, 227).

Unfortunately, that kind of visionary, long-term planning never took place. The state's population, not quite 20 million in the 1970 census, grew to 38.8 million in 2015, according to the California Department of Finance. Though the growth rate has slowed in recent decades, state planners still regularly reference projections for 50 million people by 2050 and even a future doubling of the current population.

Since World War II, California's freeway-centered lifestyle has fostered sprawling growth in a succession of "booms" that were *only* sustained because of long-distance water delivery systems. Southern California and the Bay Area overcame regional limits with imported water. Today there are about 20 million people in southern California, where local water sources might have sustained just three million. Silicon Valley, the computer industry center that fostered development across the Santa Clara Valley and regional bedroom communities, depends on water delivered via the Hetch Hetchy, Mokelumne, and California Aqueducts.

Recent commitments of water to the environment are actually not new "demands" on supply, but belated recognition that too much essential water was taken away in the past. Future increases in statewide demand predicted by water planners are based primarily on uncontrolled urban population growth. Calls for current users to conserve to stretch supplies while permitting ever more development remained, unfortunately, a familiar pattern. It would be fairest to existing water users to enact building moratoriums in times of water shortage, while working toward a long-term, sustainable relationship between Californians and their water environment.

Figure 115. Suburban sprawl in southern California. "Whoever brings the water, brings the people" (William Mulholland).

"Show me the water" laws, which require proof of a reliable water supply before approval of very large housing projects, were just taking effect as the first edition of this book appeared. These laws finally began to be applied during the 2008 drought years. Several large projects, for example, were delayed in Riverside County for a few months and moved ahead only after securing strict water-efficiency commitments from developers. Now there must be assurances of adequate water supplies before large developments (more than 500 units) can be approved. New construction should be water neutral, or mitigate its impacts through efficiency measures and infrastructure improvements in existing communities (fig. 115).

Once the benefits of a stable population capture society's attention, embracing resource limits could become the means for achieving stabilization without intervention into personal reproductive rights or the need for draconian immigration

Figure 116. Water, the lifeblood of the golden state. This photograph is of Bubbs Creek.

restrictions. Education for women and access to contraception have proven to be keys to a phenomenon occurring in at least 51 developed nations where fertility rates are now at or below replacement levels, and populations have stabilized or are even in decline. Compassionate consideration for numbers is not just important as a human issue, but also matters to every living thing in the environments we share.

We might, of course, continue the historic pattern of using water policy to facilitate growth. In the long term, then, effects on the environment and on our quality of life would worsen, despite our best efforts at habitat and species protection, and despite concepts like "smart growth." Never-ending growth, whether smart or dumb, will inevitably overtake the limits of California's water systems.

We can, instead value water for its essential role facilitating life, ours and every other living organism's. Number one on the list of 10 items in the Governor's 2014 California Water Action Plan was "Make conservation a California way of life." Our way of life should never take water for granted. Knowledge can lead to respect and a sense of wonder and awe about the Earth's most amazing molecule.

Water is the essence of life in California, as it is everywhere on this planet (fig. 116). The life we in California choose, the future we choose, will continue to be shaped most of all by decisions about water.

ACRONYMS AND ABBREVIATIONS

AF acre-foot; enough water to cover one acre to a depth of one foot; 325,851 gallons.
  1 AF historically served an average of one to two households or five to eight people per year.

APPMA American Pet Products Manufacturers Association

BDCP Bay Delta Conservation Plan

BMPS Best Management Practices

CALEPA California Environmental Protection Agency

CALFED state and federal Bay Delta Program

CEQA California Environmental Quality Act (1970)

CVP Central Valley Project (federal)

CVPIA Central Valley Project Improvement Act (1992)

CWA Clean Water Act

DFW Department of Fish & Wildlife (state; formerly Fish & Game)

DOGGR California Division of Oil, Gas, and Geothermal Resources

DWP Department of Water and Power (Los Angeles); also LADWP

DWR Department of Water Resources (state)

EBMUD East Bay Municipal Utility District

EPA Environmental Protection Agency (federal)

ESA Endangered Species Act (both state and federal)

FERC Federal Energy Regulatory Commission

IID Imperial Irrigation District

LADWP Los Angeles Department of Water and Power; also DWP

MAF million acre-feet

MTBE methyl tertiary butyl ether (gasoline additive, water pollutant)

MWD Metropolitan Water District of Southern California; also SCMWD

NODOS North of Delta Off-stream Storage

NPS nonpoint source (of water pollution)

NRDC Natural Resources Defense Council

PDO Pacific Decadal Oscillation

PG&E Pacific Gas & Electric

SDCWA San Diego County Water Authority

SFPUC San Francisco Public Utilities Commission

SWP State Water Project (since 1960, administered by DWR)

SWRCB State Water Resources Control Board

TAF thousand acre-feet

TCE trichloroethylene (industrial solvent, water pollutant)

TDS total dissolved solids (water treatment standard)

THM trihalomethanes (chlorination by-product; health risk)

TMDL total maximum daily load (specifies pollutant level allowed to meet water quality Standards)

USBR United States Bureau of Reclamation

USGS United States Geological Survey

UV ultraviolet radiation (water treatment method)

HISTORICAL TIMELINE

For more details, see Norris Hundley Jr.'s The Great Thirst: Californians and Water, 1770s–1990s *(2001). For environmental changes tied to the state's history of water development, see Carle's* Water and the California Dream: Choices for the New Millennium *(2003).*

1781 Spanish settlers erect a dam on the Los Angeles River to serve the new Pueblo de Los Angeles.

1848 Alta California becomes the property of the United States under the Treaty of Guadalupe Hidalgo, which includes language to protect pueblo water rights as identified under Mexican law.

James Marshall discovers gold at Sutter's Mill on the South Fork of the American River.

1849 Ninety thousand 49ers arrive in one year. They mine Sierra Nevada riverbeds first and later build ditches, flumes, and reservoirs to appropriate water for hydraulic mining.

Levees are constructed to try to confine waters in the San Francisco Bay–Delta.

1850 At statehood California's population is about 100,000. The Native American population has dropped to about 30,000 (down from about 300,000 before contact with Europeans).

1861 Extensive "swampland reclamation" is under way in the Central Valley, encouraged by federal and state laws.

1870 California farmers irrigate 60,000 acres.

The state's population is 560,000.

1874 A federal commission appointed by President Ulysses S. Grant proposes a storage and distribution system for the Central Valley.

1884 A federal circuit court orders hydraulic mines to stop damaging downstream property owners with runoff sediments. This order effectively stops hydraulic mining.

1887 The Wright Act becomes state law, permitting formation of irrigation districts.

1902 President Theodore Roosevelt signs the Reclamation Act. The Bureau of Reclamation begins a series of investigations on control and use of the Colorado River.

California has 2.6 million acres of irrigated farmland.

1905 The Colorado River breaks through Imperial Valley Canal headgates. After two years, the Southern Pacific Railroad is finally able to close the "leak," which has formed the Salton Sea.

Los Angeles voters authorize bonds for the Owens Valley project to bring Owens River water to Los Angeles.

1910 San Francisco voters approve the Hetch Hetchy water project to bring Tuolumne River water from Yosemite National Park.

1913 The Los Angeles Aqueduct begins delivering Owens Valley water.

1914 Water Commission Act formalizes California water rights system, granting priority to pre-1914 riparian rights over appropriative rights.

1916 In the Imperial Valley, 300,000 acres are being irrigated.

1920 The population of the city of Los Angeles reaches 576,000, surpassing that of San Francisco.

1920–23 Los Angeles buys more land and water rights in the Owens Valley.

1922 The Colorado River Compact apportions water between states.

1923 San Francisco's dam floods Hetch Hetchy Valley, though the aqueduct and tunnels will not deliver water to the city for another decade.

1924 Owens Valley residents blow up the Los Angeles Aqueduct; Owens Lake is dry due to water diversions.

1927 East Bay Municipal Utility District (EBMUD) begins building Mokelumne River aqueduct facilities.

1928 The Metropolitan Water District of Southern California (MWD) is formed; a campaign is launched to pass a bond act for a Colorado River aqueduct.

The California Constitution is amended to mandate that water not be wasted, but put to reasonable and beneficial uses.

St. Francis Dam, part of the Los Angeles water system, collapses, killing over 400 people.

1929 State, federal, and local agencies begin cooperative snowpack monitoring to forecast water supplies.

1929–34 California suffers its most severe drought since statehood. This drought becomes the standard for estimating needed storage capacity in reservoirs built later.

1930 EBMUD's Mokelumne River project is in service.

The state's population reaches 5.5 million.

1931 Southern California voters approve a $220 million bond so the MWD can begin building the Colorado River Aqueduct.

The County of Origin Law is passed, guaranteeing the right of counties to reclaim their exported water if they ever need it.

1933 Federal construction of Hoover Dam (the Boulder Canyon project) begins.

The California legislature and, later, voters in a referendum approve the Central Valley Project Act and a construction bond, but the Great Depression keeps the state from financing the project.

1934 The Hetch Hetchy Aqueduct begins delivering Tuolumne River water to San Francisco (24 years after the project is authorized).

The Bureau of Reclamation begins construction on Parker Dam, to impound water for the Colorado River Aqueduct, and on the All-American Canal, to deliver Colorado River water to the Imperial Valley.

1936 The federal Flood Control Act authorizes multipurpose dams, inaugurating an era of dam building across the West.

1937 The federal government takes over the Central Valley Project (CVP).

Los Angeles voters pass a $40 million bond to build a Mono Basin extension to the Los Angeles Aqueduct and to buy more Owens Valley land.

1938 Shasta Dam construction begins as part of the CVP. It will create the state's largest reservoir.

1940 The All-American Canal begins delivering water to the Imperial Irrigation District (IID).

The California population is 6.9 million.

1941 The MWD completes the Colorado River Aqueduct, bringing water 242 miles to Southern California.

Water diversions from Mono Lake streams begin reaching Los Angeles.

1942 Imperial Valley farms receive the first deliveries via the All-American Canal.

1943 The Mexican-American Treaty guarantees Mexico 1.5 million acre-feet per year from the Colorado River.

1945 Construction begins on Friant Dam on the San Joaquin River, as part of the CVP.

1948 Construction begins on Folsom Dam on the American River, as part of the CVP.

1950 State Attorney General Edmund G. "Pat" Brown declares that federal CVP projects need not comply with state fish protection laws, calling such releases "a waste of water."

Irrigated acreage is up to 6.5 million acres.

About 80,000 pumps are extracting groundwater in California.

The state's population is over 10.5 million.

1951 The first CVP deliveries via the Delta-Mendota Canal bring Sacramento Valley water south to the San Joaquin Valley.

1960 Voters pass a $1.75 billion bond act authorizing the State Water Project (SWP).

California has more than eight million irrigated acres.

The state's population is 15.7 million.

1962 The SWP begins construction of Oroville Dam on the Feather River.

1963 California becomes the nation's most populous state, passing New York.

1966 Construction of New Melones Dam begins on the Stanislaus River.

1968 Oroville Dam is dedicated, and the reservoir is filled to its capacity of 3.5 million acre-feet. The SWP makes its first deliveries of Northern California water to the San Joaquin Valley. The national Wild and Scenic Rivers Act becomes law.

1969 San Luis Reservoir is completed, a joint state/federal facility and the nation's largest off-stream reservoir.

1970 Owens Valley groundwater pumping and Mono Basin diversions increase to fill the second barrel of the Los Angeles Aqueduct. The National Environmental Quality Act, California Environmental Quality Act (CEQA), and California Endangered Species Act (ESA) are enacted. California is now the most urban state in the nation and still the most populous, at 20 million.

1971 The SWP's California Aqueduct begins moving northern water all the way to Southern California, pumping it nearly 2,000 feet over the Tehachapi Mountains. The California Wild and Scenic Rivers Act is passed, prohibiting new dams on North Coast rivers.

1972 The federal Clean Water Act passes, including provisions to protect wetlands. Inyo County sues Los Angeles under CEQA over groundwater and irrigation issues.

1973 The federal Endangered Species Act (ESA) is passed to protect species and critical habitats.

1974 The Safe Drinking Water Act sets federal drinking-water standards.

1975 Construction of Auburn Dam is suspended due to seismic concerns.

1976–77 California sees its driest year since record keeping began.

1976–79 A court rejects two Environmental Impact Reports (EIRs) prepared for Los Angeles's operation of a second aqueduct.

1980 Smith River and parts of the other last free-flowing rivers in Northern California are given federal Wild and Scenic Rivers status.

1982 A statewide vote rejects the Peripheral Canal, marking the first defeat of a major California water project.

1983 Deformed and dead waterfowl are found at Kesterson Reservoir; the cause is identified as toxic farm drainage.

The California Supreme Court rules that the Public Trust Doctrine applies to Mono Lake and that established water licenses must be modified when public trust values are being damaged.

1985 The Central Arizona Project comes on line; California stands to lose about 800,000 acre-feet a year of water it had been taking beyond its allowance.

1987–93 A six-year drought strikes California.

1988 The MWD and the IID sign a water transfer agreement; conservation in the Imperial Valley is to make up to 100,000 acre-feet per year available to the MWD.

1989 Winter-run Chinook Salmon *(Oncorhynchus tshawytscha)* receive emergency listing as threatened under the federal ESA and endangered under the California ESA.

1990 The state's population reaches 29.8 million.

1991 The California Drought Water Bank is created to meet short-term water needs by marketing agricultural water to cities. This measure is repeated in 1992 and 1994.

1992 The Central Valley Project Improvement Act returns 800,000 acre-feet of water to the environment each year.

A federal district court rules that state law protecting fisheries does apply to the CVP.

1993 A court rules that the CVP must obey state law and keep water for fish below dams.

The Delta Smelt *(Hypomesus transpacificus)* is declared a threatened species under the California ESA.

1994 The Bay-Delta Accord is signed, establishing the state/federal CALFED program to seek solutions to the problems of the San Francisco Bay–Delta.

The state modifies Los Angeles's water licenses to protect public trust values at Mono Lake and its tributary streams.

Winter-run Chinook Salmon are reclassified as endangered under federal law.

1995 Widespread flooding occurs in many California counties.

1997 The Coastal Branch of the SWP is completed to San Luis Obispo and Santa Barbara Counties.

Winter floods set new records along the Sacramento and San Joaquin Rivers, with damage totaling over $2 billion in 48 counties.

1998 Los Angeles and the Great Basin Air Pollution Control Board agree to resolve Owens Lake dust storms by 2006.

The IID negotiates to send 200,000 acre-feet a year to San Diego. (The controversial plan is still being debated in 2003.)

DWR's California Water Plan update estimates the state's water shortages at 1.6 million acre-feet in average water years and 5.2 million acre-feet in drought years, and forecasts increased shortages by 2020.

1999 The MWD completes construction of the off-stream Diamond Valley Reservoir near Hemet and begins filling it with Colorado River and SWP water.

Spring-run Chinook Salmon and coastal Chinook Salmon are listed as threatened under the federal and state ESAs.

After courts rule that the dewatering of the San Joaquin River violated state law, negotiations begin to develop a river restoration plan.

2000 CALFED releases a plan to address the San Francisco Bay–Delta issues.

2001 State laws mandate, for the first time, that large new developments must not be approved unless adequate water supplies are available first.

2002 When California fails to meet a year-end deadline for a plan to gradually end its reliance upon surplus Colorado River water, the Department of the Interior responds with immediate cuts for the MWD and the IID.

The state's population is about 35 million.

2003 Quantification Settlement Agreement is reached between the state's Colorado River water users to comply with the requirement for California to stop taking more than its allotted water right from the river. The agreement requires the state to take responsibility for impacts on the Salton Sea from reduced agricultural drainage in the Imperial Valley.

2005 The update to the State Water Plan, published two years late, identifies the greatest opportunities to expand water supply from conservation, recycling, and groundwater management.

2006 The governor appoints a Delta Vision Blue Ribbon Task Force to envision a sustainable future for the Sacramento–San Joaquin Delta. Its recommendations, finalized in 2008, identify coequal goals of a sustainable ecosystem and a reliable water supply.

Sixty-two miles of the lower Owens River is rewatered by court order. The stretch had been dry since Los Angeles diverted the entire flow into the Los Angeles Aqueduct in 1913.

2007 Delta pumps serving aqueducts are shut down by court order because Delta Smelt populations plummeted to near-extinction levels. New protection plans for the endangered fish species are ordered.

2008 After two drier-than-normal years, the governor declared a drought and urges the legislature to bring a new water bond before voters.

2009 "20×2020," formally the Water Conservation Act, requires 20 percent reduction in urban per capita water use in California by December 31, 2020, and agricultural water suppliers to adopt agricultural water management plans by December 31, 2012, to be updated every five years.

San Joaquin River Restoration Settlement Act required the U.S. Bureau of Reclamation to release additional flows from Friant Dam to the San Joaquin River.

Delta Reform Act creates Delta Stewardship Council as a California State Agency.

2012 The Rainwater Capture Act of 2012 and Graywater in California Plumbing Code changes were enacted to increase local capture and reuse of stormwater and drainwater from washing machines and showers.

On November 20, Minute 319 amended the 1944 treaty defining how the United States and Mexico share Colorado River water, creating a five-year pilot program to return some water to the Colorado Delta.

2013 In September, the Human Right to Water Act became California law.

2014 January 17, the Governor declared a California drought emergency.

On September 16, three bills became State law, collectively called the Sustainable Groundwater Management Act, setting standards for "sustainable yield" groundwater pumping plans.

Proposition 1, a $7.5 billion bond act funding water quality, supply, treatments, and storage projects, was approved by voters in November.

2015 As the April 1 snow survey showed a record low five percent of normal snowpack in the Sierra Nevada, Governor Jerry Brown issued the first ever mandatory water conservation directive in California history, requiring a 25 percent cut in urban water use.

AGENCIES AND ORGANIZATIONS

STATE OF CALIFORNIA

California Groundwater, www.groundwater.ca.gov

Delta Protection Commission, www.delta.ca.gov

Delta Stewardship Council, www.deltacouncil.ca.gov

Department of Conservation, www.consrv.ca.gov

Department of Fish and Game, www.dfg.ca.gov/dfghome.html

Department of Food and Agriculture, www.cdfa.ca.gov

Department of Health Services, Division of Drinking Water and Environmental Management, www.dhs.cahwnet.gov/org/ps/ddwem

Department of Water Resources, www.water.ca.gov; Water Plan update, www.waterplan/water.ca.gov; Save Our Water Campaign (with the Association of California Water Agencies), www.saveourh2o.org/content/about-save-our-water

Environmental Protection Agency (CalEPA), www.calepa.ca.gov

Reclamation Board, www.recbd.ca.gov

San Francisco Bay Conservation and Development Commission, http://ceres.ca.gov/bcdc

State Water Resources Control Board, www.swrcb.ca.gov

FEDERAL

Army Corps of Engineers, www.usace.army.mil
Bureau of Land Management, www.blm.gov/nhp/index.htm
Bureau of Reclamation, www.usbr.gov
Environmental Protection Agency, www.epa.gov
Fish and Wildlife Service, www.fws.gov
Forest Service, www.fs.fed.us
Geological Survey, http://ca.water.usgs.gov; http://sfbay.wr.usgs.gov
National Marine Fisheries Service, www.nmfs.noaa.gov
Natural Resources Conservation Service, www.ca.nrcs.usda.gov
Western Area Power Administration, www.wapa.gov

NONGOVERNMENT

A partial list

American River Conservancy, www.coloma.com/arc/index.html
Audubon California, www.audubon-ca.org
Bay Institute of San Francisco, www.bay.org
California League of Conservation Voters, www.ecovote.org/ecovote
California Wild Heritage Campaign, www.californiawild.org
CalTrout, www.caltrout.org
Environmental Defense Fund, www.environmentaldefense.org
Friends of the Los Angeles River, www.folar.org
Friends of the River, www.friendsoftheriver.org
Friends of the Trinity River, www.fotr.org
Heal the Bay, www.healthebay.org
Maven's Notebook, www.mavensnoteook.com
Mono Lake Committee, www.monolake.org
Natural Resources Defense Council, www.nrdc.org
Owens Valley Committee, www.ovcweb.org
Restore Hetch Hetchy, www.hetchhetchy.org
South Yuba River Citizens League, www.syrcl.org
Santa Monica BayKeeper, www.smbaykeeper.org

Sierra Club, www.sierraclub.org

SoCal Water$mart, residential and commercial rebate programs through MWD, www.socalwatersmart.com

TreePeople, www.treepeople.org

WATER RESOURCE INSTITUTES AND RESEARCH CENTERS

Center for Water Resources, UC Riverside, www.waterresources.ucr.edu

Pacific Institute, www.pacinst.org

Water Education Foundation, www.water-ed.org

Water Resources Center Archives, UC Berkeley, www.lib.berkeley.edu/WRCA

REFERENCES

American Farmland Trust. 2005. *The Future Is Now: Central Valley Farmland at the Tipping Point?* Washington, DC: American Farmland Trust. Available at www.farmland.org/programs/states/futureisnow/. Accessed January 26, 2015.

Arthus-Bertrand, Yann, David Baker, Lester Russell Brown, and Jean-Marie Pelt. 1999. *Earth from Above.* New York: Harry N. Abrams.

Bakker, Elna. 1984. *An Island Called California.* Berkeley: University of California Press.

Ball, Philip. 2001. *Life's Matrix: A Biography of Water.* Berkeley: University of California Press.

Barlow, Maude, and Tony Clarke. 2002. *Blue Gold: The Fight to Stop the Corporate Theft of the World's Water.* New York: New Press.

Barnett, Tim P., and David W. Pierce. 2008. When will Lake Mead go dry? *Water Resources Research* 44:W03201, doi:10.1029/2007WR006704.

Bay Institute of San Francisco. 1998. *From the Sierra to the Sea: The Ecological History of the San Francisco Bay–Delta Watershed.* San Rafael, CA: The Bay Institute of San Francisco.

Borchers, James W., and Michael Carpenter. 2014. *Land Subsidence from Groundwater Use in California.* Sacramento: California Water Foundation.

Bryant, Edwin. [1848] 1985. *What I Saw in California*. Lincoln: University of Nebraska Press.

CALFED Bay-Delta Program. 1998. *Executive Summary: Draft Programmatic Environmental Impact Statement/Environmental Impact Report*. Sacramento.

———. 2002. *CALFED Bay-Delta Program Annual Report, 2001*. Revised. Sacramento.

California Dams Database, Berkeley Digital Library Project. n.d. Available at http://elib.cs.berkeley.edu/dams/dam-map.html.

California Department of Conservation. 2000. *Farmland Conversion Report, 1998–2000*. Sacramento.

California Department of Fish and Wildlife. n.d. Sacramento. Available at www.dfg.ca.gov/fish/hatcheries. Accessed June 2, 2015.

California Department of Water Resources. 1998. *Bulletin 160–98: California Water Plan*. Sacramento.

———. 2010. *20×2020 Water Conservation Plan*. Sacramento. Available at www.swrcb.ca.gov/water_issues/hot_topics/20x2020. Accessed January 2, 2015.

———. 2013a. *Bay Delta Conservation Plan Highlights*. Sacramento. Available at http://baydeltaconservationplan.com/PublicReview/InformationalMaterials. Accessed December 28, 2014.

———. 2013b. *California Water Plan: Update 2013*. Sacramento. Available at www.waterplan.water.ca.gov/. Accessed December 29, 2014.

———. 2013c. Salton Sea Species Conservation Habitat Project – Final EIS/EIR and Final Design updates. Available at www.water.ca.gov/saltonsea/#eir. Accessed December 29, 2014.

———. 2014a. *Groundwater Information Center*. Available at www.water.ca.gov/groundwater/data_and_monitoring/groundwater_reports.cfm. Accessed January 6, 2015.

———. 2014b. *Historical Sacramento and San Joaquin valley water year type index*. Available at http://cdec.water.ca.gov/cgi-progs/iodir/WSIHIST. Accessed January 14, 2015.

California Natural Resources Agency. 2008. *Our Vision for the California Delta: Governor's Delta Vision Blue Ribbon Task Force. Final report*. Sacramento. Available at http://deltavision.ca.gov/index.shtml. Accessed January 1, 2015.

———. 2014. *California Water Action Plan*. Sacramento. Available at http://resources.ca.gov/california_water_action_plan/. Accessed January 19, 2015.

California State Water Resources Control Board. 1997. *California's Rivers and Streams: Working toward Solutions*. Available at http://www.swrcb.ca.gov/riversst.htm.

———. 1999. *A Guide to Water Transfers*. Sacramento: State Water Resources Control Board, Division of Water Rights, and California Environmental Protection Agency.

———. 2001. *Opportunity, Responsibility, Accountability: Nonpoint Source Pollution Control Program*. Sacramento: State Water Resources Control Board and California Environmental Protection Agency.

———. 2010. *Final Report on the Development of Flow Criteria for the Sacramento-San Joaquin Delta Ecosystem*. Sacramento: State Water Resources Control Board. Available at www.waterboards.ca.gov/waterrights/water_issues/programs/bay_delta/deltaflow/index.shtml. Accessed January 21, 2015.

California Water Impact Network. 2013. *Principles of agreement for a proposed settlement between the United States and Westlands Water District regarding drainage*. California Water Impact Network. Available at www.c-win.org/webfm_send/453. Accessed January 11, 2015.

Cantu, Celeste. 2013. *A 21st Century Relationship with Water*. TEDx Temecula. Available at https://www.youtube.com/watch?v=StUyp2ioVXs&feature=youtu.be. Accessed January 1, 2015.

Carle, David. 2003. *Water and the California Dream: Choices for the New Millennium*. San Francisco: Sierra Club Books (Paperback edition. 2000. *Drowning the Dream: California's Water Choices at the Millennium*. Westport, CT: Praeger).

———. [2004] 2009. *Introduction to Water in California*. Berkeley: University of California Press.

———. 2006. *Introduction to Air in California*. Berkeley: University of California Press.

———. 2008. *Introduction to Fire in California*. Berkeley: University of California Press.

———. 2010. *Introduction to Earth, Soil, and Land in California*. Berkeley: University of California Press.

Carle, David and Janet Carle. 2013. *Traveling the 38th Parallel: A Water Line around the World*. Berkeley: University of California Press.

Carson, James H. [1852] 1931. *Early Recollections of the Mines, and a Description of the Great Tulare Valley...Steamer Edition of the San Joaquin Republican*. Reprint, New York: W. Abatt.

Cayan, Dan, Mary Tyree, Mike Dettinger, Hugo Hidalgo, Tapash Das, Ed Maurer, Peter Bromirski, Nicholas Graham, and Reinhard Flick. 2009. *Climate Change Scenarios and Sea Level Rise Estimates for the California 2008 Climate Change Scenarios Assessment*. CEC-500-2009-014-D. Report Prepared by the California Climate Change Center for the California Energy Commission. Sacramento.

Childs, Craig. 2000. *The Secret Knowledge of Water*. Seattle: Sasquatch Books.

Circle of Blue. 2014. *'Transformational' Water Reforms, though Wrenching, Helped Australia Endure Historic Drought, Experts Say*. Available at www.circleofblue.org/waternews/2014/world/transformational-water-reforms-though-wrenching-helped-australia-endure-historic-drought-experts-say. Accessed March 19, 2014.

Cohen, Andrew. 1994. The hidden costs of California's water. In *Life on the Edge: A Guide to California's Endangered Natural Resources*, edited by Carl G. Thelander, 288–302, Berkeley: BioSystems Books.

Cohen, Michael J., Jason I. Morrison, and Edward P. Glenn. 1999. *Haven or Hazard: The Ecology and Future of the Salton Sea*. Oakland: Pacific Institute for Studies in Development, Environment and Security.

Concerned Health Professionals of New York. 2014. *Compendium of Scientific, Medical, and Media Findings Demonstrating Risks and Harms of Fracking (Unconventional Gas and Oil Extraction)*. 2nd ed. Available at http://concernedhealthny.org/compendium. Accessed December 13, 2014.

Cooley, Heather. 2014. *Multiple Benefits of Water Conservation and Efficiency for California Agriculture*. Oakland: Pacific Institute. Available at http://pacinst.org/publication/multiple-benefits-of-water-conservation-and-efficiency-for-california-agriculture/. Accessed January 20, 2015.

Cooley, Heather, Newsha Ajami, and Matthew Heberger. 2013. *Key Issues in Seawater Desalination in California: Marine Impacts*. Oakland: Pacific Institute. Available at www.pacinst.org/publication/desal-marine-impacts. Accessed January 17, 2015.

Cooley, Heather, Peter H. Gleick, and Gary Wolff. 2006. *Desalination with a Grain of Salt: A California Perspective*. Oakland: Pacific Institute. Available at www.pacinst.org/reports/desalination/desalination_report.pdf. Accessed July 7, 2008.

Cox, John. 2014. State poised to shut down 11 local oil injection wells. *Bakersfield Californian*. July 3. Available at www.bakersfieldcalifornian.com/business/kern-gusher/x634489929/State-poised-to-shut-down-11-local-oil-injection-wells. Accessed July 5, 2014.

Daniels, Tom, and Deborah Bowers. 1997. *Holding Our Ground: Protecting America's Farms and Farmland*. Washington, DC: Island Press.

Delta Stewardship Council. 2013. *Delta Plan*. Available at www.deltacouncil.ca.gov/delta-plan-0. Accessed January 4, 2015.

Didion, Joan. 1979. *The White Album*. New York: Farrar, Straus & Giroux.

Dinno, Rachel, ed. 1999. *Restoring the California Dream: Ten Steps to Improve Our Quality of Life*. Sacramento: Planning and Conservation League Foundation.

Duncan, David James. 2001. *My Story as Told by Water*. San Francisco: Sierra Club Books.

Durrenberger, Robert W. 1968. *Patterns on the Land*. Palo Alto, CA: National Press Books.

Eiseley, Loren. 1962. *The Immense Journey*. Chicago: Time-Life Books.

Field, Christopher B., Gretchen C. Daily, Frank W. Davis, Steven Gaines, Pamela A. Matson, John Melack, and Norman L. Miller. 1999. *Confronting Climate Change in California*. Cambridge, MA: Union of Concerned Scientists; Washington, DC: Ecological Society of America.

Friends of the River. 1999. *Rivers Reborn: Removing Dams and Restoring Rivers in California*. Sacramento: Friends of the River.

Gleick, Peter H. 2011. *Bottled and Sold: The Story Behind Our Obsession with Bottled Water*. Washington, DC: Island Press.

Green, Dorothy. 2007. *Managing Water: Avoiding Crisis in California*. Berkeley: University of California Press.

Griffin, Griffin, and Kevin J. Anchukaitis. 2014. How unusual is the 2012 to 2014 California drought? *Geophysical Research Letters*, 41: 9017–23. doi: 10.1002/2014GL062433.

Hall, Clarence A. Jr., Victoria Doyle-Jones, and Barbara Widawski, eds. 1992. *The History of Water: Eastern Sierra Nevada, Owens Valley, White-Inyo Mountains*. Vol. 4, *White Mountain Research Station Symposium*. San Francisco: Regents of the University of California.

Hanak, Ellen, and Elizabeth Stryjewski. 2012. *California's Water Market, by the Numbers: Update 2012*. San Francisco: Public Policy Institute of California.

Hanson, Warren D. [1985] 1994. *San Francisco Water and Power*. San Francisco: City and County of San Francisco.

Hart, John. 1996. *Storm over Mono: The Mono Lake Battle and the California Water Future*. Berkeley: University of California Press.

Hicks, Tom, and Alf W. Brandt. 2013. *Layperson's Guide to Water Rights Law*. Sacramento: Water Education Foundation.

Hoffman, Abraham. 1981. *Vision or Villainy? Origins of the Owens Valley–Los Angeles Water Controversy*. College Station: Texas A&M University Press.

Holder, Charles Frederick. 1906. *Life in the Open: Sport with Rod, Gun, Horse and Hound in Southern California*. New York: Putnam's.

Hundley, Norris Jr. 2001. *The Great Thirst: Californians and Water, 1770s–1990s*. Berkeley: University of California Press.

Ingram, B. Lynn, and Frances Malamud-Roam. 2013. *The West Without Water: What Past Floods, Droughts, and Other Climatic Clues Tell Us about Tomorrow*. Berkeley: University of California Press.

Intergovernmental Panel on Climate Change. 2014. *Fifth Assessment Report*. Geneva: Intergovernmental Panel on Climate Change. Available at www.ipcc.ch/. Accessed December 29, 2014.

International Boundary and Water Commission, United States and Mexico. 2012. *Minute No. 319*. Available at www.ibwc.state.gov/Treaties _Minutes/Minutes.html. Accessed January 7, 2015.

Kahrl, William L. 1979. *The California Water Atlas*. Sacramento: California Governor's Office of Planning and Research.

————. 1982. *Water and Power.* Berkeley: University of California Press.

Kattelmann, Richard. 1996. Hydrology and water resources. In *Status of the Sierra Nevada: Sierra Nevada Ecosystem Project.* Vol. 2, *First report to Congress.* Davis: University of California, Davis, Centers for Water and Wildland Resources.

Kelley, Robert. 1989. *Battling the Inland Sea.* Berkeley: University of California Press.

Kolpin, Dana W., Edward T. Furlong, Michael T. Meyer, E. Michael Thurman, Steven D. Zaugg, Larry B. Barber, and Herbert T. Buxton. 2002. Pharmaceuticals, hormones, and other organic wastewater contaminants in U.S. streams, 1999–2000: a national reconnaissance. *Environmental Science Technology* 36 (6): 1202–11.

Lehner, Peter, George P. Aponte Clark, Diane M. Cameron, and Andrew G. Frank. 1999. *Stormwater Strategies: Community Responses to Runoff Pollution.* New York: Natural Resources Defense Council.

Lemly, A. Dennis. 1997. Environmental implications of excessive selenium: a review. *Biomedical and Environmental Sciences* 10: 415–35.

Leopold, Aldo. [1966] 1974. *A Sand County Almanac.* New York: Sierra Club/Ballantine Books.

Leopold, A. Starker. 1984. *Wild California: Vanishing Land, Vanishing Wildlife.* Berkeley: University of California Press.

Los Angeles Department of Water and Power. 1988. *Sharing the Vision: The Story of the L.A. Aqueduct.* Los Angeles.

Ludwig, Art. 2012. *Create an Oasis with Greywater.* 5th ed. Santa Barbara, CA: Oasis Design.

Lufkin, Alan. 1991. *California's Salmon and Steelhead: The Struggle to Restore an Imperiled Resource.* Berkeley: University of California Press.

Lustgarten, A. (2014). California halts injects of fracking waste, warning it may be contaminating aquifers. *ProPublica.* Available at www.propublica.org/article/ca-halts-injection-fracking-waste-warning-may-be-contaminating-aquifers. Accessed January 5, 2015.

Madgic, Bob. 2013. *The Sacramento: A Transcendent River.* Anderson, CA: Riverbend Books.

Mahoney, Laura. 2001. *Layperson's Guide to Agricultural Drainage.* Sacramento: Water Education Foundation.

Mayer, Jim. 1996. *Layperson's Guide to Water Pollution.* Sacramento: Water Education Foundation.

McClurg, Sue. 1998. Saving the salmon. *Western Water* (January/February): 4–13.

―――. 2000a. The Mojave River basin decision. *Western Water* (September/October).

―――. 2000b. *Water and the Shaping of California.* Sacramento: Water Education Foundation.

―――. 2001. Conjunctive use: banking for a dry day. *Western Water* (July/August).

Metropolitan Water District of Southern California. 1997. *Wheeling: Gearing for the Future of Water Marketing.* Los Angeles.

Miller, M. A., I. A. Gardner, C. Kreuder, D. M. Paradies, K. R. Worcester, D. A. Jessup, E. Dodd, M. D. Harris, J. A. Ames, A. E. Packham, and P. A. Conrad. 2002. Coastal freshwater runoff is a risk factor for *Toxoplasma gondii* infection of southern sea otters *(Enhydra lutris nereis). International Journal for Parasitology* 32 (8): 997–1006.

Montgomery, Gayle B. 1999. *Its Name Was M.U.D., Book II.* Oakland: East Bay Municipal Utility District.

Mount, Jeffrey F. 1995. *California Rivers and Streams.* Berkeley: University of California Press.

Muir, John. [1901] 1991. *Our National Parks.* San Francisco: Sierra Club Books.

―――. 1912. *The Yosemite.* New York: Century Company.

―――. [1938] 1979. *John of the Mountains: The Unpublished Journals of John Muir,* edited by Linnie Marsh Wolfe. Madison: University of Wisconsin Press.

―――. 1961. *The Mountains of California.* New York: Doubleday/American Museum of Natural History.

National Academy of Science. 2006. *Flouride in Drinking Water.* Washington, DC. http://dels.nas.edu/Materials/Report-In-Brief/4775-Fluoride. Accessed January 7, 2015.

Office of Environmental Health Hazard Assessment, California Environmental Protection Agency. 2013. *Indicators of Climate Change in California, August 2013.* Available at www.oehha.ca.gov/multimedia/epic/2013EnvIndicatorReport. Accessed January 17, 2015.

Pacific Institute. 2007. *Bottled Water and Energy Fact Sheet.* Available at http://pacinst.org/publication/bottled-water-and-energy-a -fact-sheet/. Accessed January 26, 2015.

Pacific Institute and the Natural Resources Defense Council. 2014. *Urban Water Conservation and Efficiency – Enormous Potential Close to Home.* Oakland. Available at http://pacinst.org/nrdc-switchboard-urban-water-conservation-and-efficiency-enormous-potential-close-to-home/. Accessed January 26, 2015.

Perrone, Debra, and Melissa Rohde. 2014. *Storing Water in California: What Can $2.7 Billion Buy Us?* Stanford: Stanford Woods Institute for the Environment. Available at http://waterinthewest.stanford.edu /resources/publications-directory. Accessed January 23, 2015.

Postel, Sandra. 2014. *With Water, Life Returns to the Colorado Delta.* December 19. Washington, DC: National Geographic Society. http:// voices.nationalgeographic.com/2014/12/19/with-water-life-returns-to-the-colorado-river-delta/. Accessed January 8, 2015.

Powell, James Lawrence. 2008. *Dead Pool, Lake Powell, Global Warming, and the Future of Water in the West.* Berkeley: University of California Press.

Reisner, Marc. 1986. *Cadillac Desert: The American West and Its Disappearing Water.* New York: Viking Penguin.

———. 1997. *Water Policy and Farmland Protection: A New Approach to Saving California's Best Agricultural Lands.* American Farmland Trust. Available at www.farmland.org/Farmland/files/water/cff.htm.

Rockwell, Robert L. 2002. *Giardia lamblia* and giardiasis, with particular attention to the Sierra Nevada. *Sierra Nature Notes.* Available at http://www.yosemite.org/naturenotes/Giardia.htm.

Rose, Gene. [1992] 2000. *The San Joaquin: A River Betrayed.* Clovis, CA: World Dancer Press.

San Francisco Public Utilities Commission. n.d. *Hetch Hetchy Regional Water System.* Available at sfwater.org/detail.cfm/MC_ID/5/MSC _ID/52/MTO_ID/NULL/C_ID/555.

Santa Barbara, City of. n.d. *Water supply sources/desalination.* Available at http://ci.santa-barbara.ca.us/departments/public_works/water _resources/bfsupply.html#Desal.

Schoenherr, Allan A. 1992. *A Natural History of California.* Berkeley: University of California Press.

ScienceDaily. 1999. Pets may be major cause of water pollution in urban areas. *ScienceDaily Magazine*, December 6. Available at http://www.sciencedaily.com/releases/1999/12/991206071651.htm.

Seeley, Rachael. 2014. Monterey proves more complex than average shale play. *UOGR*, April 13. Available at www.ogj.com/articles/uogr/print/volume-2/issue-4/monterey-proves-more-complex-than-average-shale-play.html. Accessed January 5, 2015.

Shigley, Paul. 2002. Cities pressure San Francisco to repair Hetch Hetchy. *California Planning and Development Report* 17 (4). Archived at http://www.cp-dr.com.

SMRC. n.d. *Pollution prevention fact sheet: animal waste collection.* Available under "Fact sheets-pollution prevention-residential prevention practices-animal waste collection," at www.stormwatercenter.net. Accessed January 27, 2015.

Snyder, Gary. 1995. *A Place in Space.* Washington, DC: Counterpoint.

Spriggs, Elisabeth Mathieu. 1931. *The History of the Domestic Water Supply of Los Angeles.* Master's thesis, University of Southern California.

Srebotnjak, Tanya, and Miriam Rotkin-Ellman. 2014. *Drilling in California: Who's at Risk?* Los Angeles: Natural Resources Defense Council.

Stegner, Wallace. 1985. *The Sound of Mountain Water.* Lincoln: University of Nebraska Press.

———. 1992. *Where the Bluebird Sings to the Lemonade Springs.* New York: Random House.

Steinbeck, John. [1952] 1995. *East of Eden.* New York: Viking Penguin.

Stene, Eric A. 1998. *The Central Valley Project.* U.S. Bureau of Reclamation. Available at www.usbr.gov/history/cvpintro.htm.

Stine, Scott. 1994. Extreme and persistent drought in California and Patagonia during mediaeval time. *Nature* 369: 546–49.

Storer, Tracy I., and Robert L. Usinger. 1963. *Sierra Nevada Natural History.* Berkeley: University of California Press.

Svete, Stephen. 2002. Cheaper, better desalination gets a fresh look. *California Planning and Development Report* 17 (2) (February). Archived at http://www.cp-dr.com.

Taylor, Mac. 2015. *Achieving State Goals for the Sacramento-San Joaquin Delta.* Sacramento: Legislative Analyst's Office. Available at http://

lao.ca.gov/reports/2015/res/Delta/sac-sj-delta-011515.aspx. Accessed January 16, 2015.

Thornton, Joe. 2000. *Pandora's Poison: Chlorine, Health, and a New Environmental Strategy.* Cambridge: MIT Press.

Twain, Mark. [1872] 1972. *Roughing It.* Berkeley: University of California Press.

University of Redlands. 2002. *Salton Sea Atlas.* Redlands, CA: Redlands Institute, University of Redlands.

U.S. Army Corps of Engineers, Sacramento District. 2002. *American River Watershed, California: Long-Term Study. Final Supplemental Plan, Formulation Report/Environmental Impact Statement/Environmental Impact Report.* Sacramento.

————. 2014. *Folsom Dam Raise Project: Public Scoping Meetings Summary Report.* Sacramento.

U.S. Bureau of Reclamation. 1987. *Hetch Hetchy: A Survey of Water and Power Replacement Concepts, Prepared on Behalf of the National Park Service by the Bureau of Reclamation, Mid Pacific Region.* Draft. Sacramento.

————. 1999. *Central Valley Project Improvement Act: Final Programmatic Environmental Impact Statement.* Sacramento.

————. 2000. *The Salton Sea Restoration Project: Opportunities and Challenges.* Washington, DC.

————. 2011. *Shasta Lake Water Resources Investigation. Draft Feasibility Report.* Sacramento.

————. 2014. *Upper San Joaquin River Basin Storage Investigation. Draft Feasibility Report.* Sacramento.

U.S. Environmental Protection Agency. 2000. *Methyl Bromide Phase Out.* Washington, DC.

U.S. Environmental Protection Agency, Region 9. 1999. *California Dairy Quality Assurance Program Fact Sheet, September 1999.* Available at www.epa.gov/region09/cross_pr/animalwaste/dairyfact.html.

U.S. Fish and Wildlife Service. 2008. *Long-Term Operational Criteria and Plan (OCAP) for Coordination of the Central Valley Project and State Water Project.* Available at www.fws.gov/sfbaydelta/cvp-swp/cvp-swp.cfm. Accessed December 20, 2014.

U.S. Geological Survey. n.d. *Access USGS-San Francisco Bay and Delta Team.* Available at http://sfbay.wr.usgs.gov/access/Integrated-Science/IntSci.html.

Versluis, Arthur. 2001. The waters under the earth. In *Writing on Water,* edited by David Rothenberg and Marta Ulvaeus. 81–88. Cambridge: Massachusetts Institute of Technology, Terra Nova.

Walton, John. 1992. *Western Times and Water Wars.* Berkeley: University of California Press.

Warren, Earl. 1977. *The Memoirs of Chief Justice Earl Warren.* Garden City, NY: Doubleday.

Warshall, Peter. 1994. Streaming wisdom: watershed consciousness in the twenty-first century. *River Voices.* Portland: River Network. Summer. http://csf.Colorado.edu/bcwatershed/BNC1a.htm.

Water Education Foundation. 1991. *Colorado River Water Map.* Sacramento: Water Education Foundation.

———. 2008. *Layperson's Guide to California Water.* Sacramento: Water Education Foundation.

———. 2014. *Where Does My Water Come from?* Sacramento: Water Education Foundation. Available at www.watereducation.org/where-does-my-water-come. Accessed January 27, 2015.

Worster, Donald. 1985. *Rivers of Empire: Water, Aridity, and the Growth of the American West.* New York: Pantheon.

Yoshiyama, Ronald M., Eric R. Gerstung, Frank W. Fisher, and Peter B. Moyle. 1996. Historical and present distribution of Chinook Salmon in the Central Valley drainage of California. In *Status of the Sierra Nevada: Sierra Nevada Ecosystem Project.* Vol. 3, *Final Report to Congress.* Davis: University of California, Davis, Centers for Water and Wildland Resources.

Zetland, David. 2014. *Living with Water Scarcity.* Amsterdam: Aguanomics Press.

PHOTO CREDITS

All photographs (figure numbers) are by the author except as noted below.

15, 16: Photos by Richard Kattelmann.

26, 29, 35, 38, 81, 95: Photos by Frank Balthis.

20: Photo courtesy of NASA.

19, 21, 24, 28, 37, 40, 46, 50, 53-59, 65, 75, 76, 78–80, 107: Photos courtesy of California Department of Water Resources.

36, 41: Photos courtesy of U.S. Fish and Wildlife Service.

48, 68, 96, 102, 104: Photos courtesy of Mono Lake Committee (photo for fig. 48 by Bartshe Miller, and photos for figs. 102 and 104 by Herley Jim Bowling).

43: Photo by Forbes, courtesy of County of Inyo, Eastern California Museum.

45: Photo courtesy of San Bernardino County Museum.

51: Photo courtesy of California State Parks.

60: Photo courtesy of U.S. Bureau of Reclamation.

69: Photo courtesy of Yosemite Museum, Yosemite National Park.

85a, b: Photos courtesy of Joe Skorupa, U.S. Fish and Wildlife Service.

88: Photo courtesy of U.S. Bureau of Reclamation.

100: Photo courtesy of John Mott.

101: Photo courtesy of TreePeople.

108: Photo by Rita Schmidt Sudman, Water Education Foundation.

111: Photo courtesy of Los Angeles Department of Water and Power.

113: NASA Space Shuttle image, June 2002.

114: Photo by Yann Arthus-Bertrand (Bertrand et al. 1999).

Author photo: Ryan Carle.

INDEX

acetaminophen, 189
acid rain, 21
acidic mine drainage, in Sacramento River, 183(f)
A.D. Edmonston Pumping Plant, 103, 104(f)
age dating, of tree stumps, 30
agribusinesses, 212
agricultural drainage, 114, 171
agricultural land, 32, 160
agricultural water suppliers, 236, 239, 288
air cooling and condensing, 6(f)
air pollution, 125, 176, 184
air quality, 80, 125, 263(f)
    monitoring of, 188
algae blooms, 174
Alkali Flies *(Ephydra hians)*, 78, 209
Amargosa River, 80
Amargosa Vole *(Microtus californicus scirpensis)*, 80
American Association of Public Health, 197
American Bird Conservancy, 77, 78

American Dental Association, 197
American Medical Association, 197
American Pet Products Manufacturers Association, 202
    National Pet Owners Survey, 202
American River, 62, 90, 108, 111, 131–32, 248, 251–52
    Salmon Festival, 158(f)
    salmon fishermen and rafters, 63(f)
Ansel Adams Wilderness, 70
Antelope Creek, 62
Antelope Dam and reservoir, 99(f)
Antelope Lake, 98
Antelope Valley, 105
aqueduct systems, xiv–xv, 40
    California, 118, 179, 211, 243
    Colorado River, 118, 119(f)
    Los Angeles, 122(map), 123–26, 201
    South Bay, 179
aquifers, xiii, 15–16, 50, 52, 54, 179, 181–82, 185, 187, 225, 227, 241

Army Corps of Engineers, 84, 108, 251

Arroyo Seco River, 73

arsenic, 182

Artesian well, near San Bernardino, 83(f)

Association of California Water Agencies, 236

atmospheric
  pressure, 2
  rivers, 32
  vapor, 21

Auburn Dam, 111, 248–49
  construction of, 285

Babbitt, Bruce, 253, 254(f)

Ballona Creek, 84

balloon-voyages, 64

Ball, Philip, 15

"Ban the Bottle" campaigns, 197

Battle Creek, 61

Bay-Delta Accord, 212

Bay Delta Conservation Plan (BDCP), 213, 215, 218, 220

Bear River, 62

Beaver (*Castor canadensis*), 46

Berryessa Lake, 133

Big Chico Creek, 62

bigcone spruce (*Pseudotsuga macrocarpa*), 29–30

Big Sur River, 73

biomagnification, 168

biomass, 153

birth-control hormones, 189

Black-neck Stilt (*Himantopus mexicanus*), 168(f)

Black Rail (*Laterallus jamaicensis*), 173

Board of Directors of Orange County Sanitation District (OCSD), 192

bottled-water phenomenon, 194–97, 195(f)

Brannan, Sam, 62

Bridgeport Valley, 64, 66

Brown, Edmund G. (Pat), 217, 273

Brown, Jerry, 216–17

Brown Pelican (*Pelecanus occidentalis*), 173

Bryant, Edwin, 60

Bubbs Creek, 276(f)

Buena Vista Lake, 37, 75

Butte Creek, 62, 154, 253

Cadiz Corporation, 244

Cadiz Valley Water Conservation, Recovery and Storage Project, 244

Calaveras Big Trees State Park, 71

Calaveras Reservoir, 256

CALFED Science Program. *See* Delta Science Program

California
  aqueduct, xiii, 91, 100, 101(f), 105, 106(f), 111, 118, 179, 211, 243, 274
  average annual precipitation, 3(map), 94(f)
  average annual streamflow, 10(map)
  average temperature, 144(f)
  Coastal Commission, 60
  Cooperative Snow Surveys Program, 29
  Department of Fish and Wildlife, 154, 156(f)
  Department of Health Services, 193
  Department of Water Resources (DWR), 29
  desalination facilities at, 267(map)
  Division of Oil, Gas, and Geothermal Resources (DOGGR), 187

Eco Restore, 220
Energy Commission, 142, 253
epic drought periods during
    Middle Ages, 30
Farm Bureau, 159
groundwater basins, 53(map)
High Speed Rail, 179
Monterey Shale formation, 185,
    186(map)
oil and gas fields in, 186(map)
Plumbing Code, 208, 230
provisions against unreasonable
    use of water, 208
suburban sprawl in, 275(f)
State Water Project allocations,
    98(t)
stormwater capture and
    graywater reuse, 227–31
Urban Water Conservation
    Council, 235
Water Code, 206, 259
Water Commission, 221, 245
water delivery project in, 123
Water Fix plan, 214(map), 220
water law and public trust,
    206–10
waterscape, xv, xvi
Water Service Company, 245
Wetlands Conservation Policy,
    207
wild and scenic rivers of, 48,
    49(map), 50
California dams, 148(map)
"California First Days" celebra-
    tion, 273
California Freshwater Shrimp
    (Syncaris pacifica), 67
California Gulls *(Larus californicus)*,
    209
California Water Plan (1957), 203,
    239, 277
CalTrout, 210, 247

Camanche Reservoir, 131
carbon dioxide, xii, 13, 16, 43,
    145–46, 197
atmospheric, 141, 142(f)
and global warming, 142
carbon footprint, 146
carcinogens, 182, 184, 192
Carmel River, 73, 134, 254
Carson, James, 75
Carson River, 66
Castaic Lake, 105
Castaic Pumping Plant, 105(f)
Cedar Creek, 84
cellular metabolism, 18
Centers for Disease Control
    (CDC), 197
Central Coast, 55, 73–74
Central Valley Project (CVP),
    California, xvii, 41, 94, 100,
    108–15, 159, 169, 206
Central Valley Project Improve-
    ment Act (CVPIA), 112
for controlling seasonal
    flooding, 108
environmental consequences of,
    113
exchange contractors, 114
facilities, 109(map)
Improvement Act (1992), 207
land grants, 112
operation of, 112
primary purposes of, 112
water rights, 112, 206
wetlands, 46
Central Valley Project Improve-
    ment Act (CVPIA), 259
Chino Basin, 164, 261
chloramines, 193
chlorination
    health problem due to exposure
        to, 193
    for water cleaning, 192

chromium-6, 183–84
City of Long Beach, 223
clean water, 260–62
Clean Water Act (CWA), 166, 207, 262, 266
Clear Lake, 63
climate change, 28, 272–73
    effects on California's water systems, xvi, 143
    and water cycle, 141–46
climate of California, 24–36
    droughts, 26–31
    floods, 31–36
Coachella Valley, 121
Coachella Valley Water District (CVWD), 117
coastal estuaries, 46
coastal marshes, of California, 44, 45(f)
Coho salmon *(Oncorhynchus kisutch)*, 57
Coleman Fish Hatchery, 110
Colorado Delta, 119
Colorado Desert, 6
Colorado River, 8, 40–41, 55, 85–87, 91, 94, 120, 147, 172–77, 244, 258
    aqueduct, 118
    conservation program for, 172
    contamination of, 183
    fish populations, 86–87
    at Picacho State Park, 86(f)
    Quantification Settlement Agreement (2003), 173
    salinity issues, 87
    service areas, 117(map)
    water delivery systems, 115–21
    water rights, 172–73
    watershed and facilities, 116(map)
    Water Use Plan (4.4 Plan), 120
Common Loons *(Gavia immer)*, 67
community water fluoridation, 197

Convict Lake, 80
Corning Canal, 110
Cosumnes Nature Preserve, 70
Cosumnes River, 48
cotton harvest, in the Tulare Basin, 76(f)
Council for Watershed Health, 228
crop production, 238
crops, agricultural
    rice fields, 165
    tolerances for soil salinity, 166, 170
    water availability for, 160
Cross Valley Canal, 242
Cryptosporidium parasite, 195
CVP Dam, 248
CVP Improvement Act (1992), 207

dairy products, 163
dams
    construction of, 151, 242, 247–49
    debate over, 245–47
    disadvantages of, 147
    hydraulic, 43
    off-stream, 249–50
    raising of, 251–52
    razing of, 252–57
    removal of, 254
    uses of, 147
    water storage capacity, 255
Davis Lake, 98
Death Valley, 8, 78, 80
Deer Creek, 62
Delta marshlands, 41–42
Delta–Mendota Canal, 100, 111, 113
    pumping from the Delta into, 111(f)
Delta Science Program, 212–13
Delta Smelt *(Hypomesus transpacificus)*, 43, 211, 216, 218
Delta Stewardship Council, 212
Delta Vision Blue Ribbon Task Force, 212

dental fluorosis, 198
Department of Water Resources
    (DWR), 95, 173, 211
desalination of water, 220, 269(f)
desalination plants, 266, 268–71
    in California, 267(map)
    in Kuwait, 269(f)
    solar powered, 170
    waste from, 269(f)
desert wildlife and plants, 80
"designer" water, 226
Devil's Postpile National Monu-
    ment, 70
Diamond Valley Reservoir, 118,
    249, 250(f)
Diazinon (dormant-spray
    pesticide), 166
disease-causing microbes, 193
distilled water, 170, 225
Doheny State Beach lagoon, 190(f)
domestic water systems, 199
Donner Summit, 4
Dos Amigos Pumping Plant,
    California, 102(f)
Douglas-firs *(Pseudotsuga menziesii)*,
    55
"Drains to Bay" message, 191(f)
drinking water systems, 223
drinking water utilities, 193
drip irrigation, 160, 170
    of grapes, 161(f)
droughts, xvi, xvii, 2, 24, 26–31, 95,
    220, 223, 231, 272
    dry cycles, 30
    due to population growth, 29
    in Sacramento and San Joaquin
        valleys, 28(t)
    sign of, 25(f)
    wet cycles, 30
Drought Water Bank, 257
    creation of, 286
dust storms, 80, 262, 263(f), 287

Eared Grebes *(Podiceps nigricollis)*,
    173
earthquakes, 35, 118, 130, 248–49
East Bay MUD. *See* East Bay
    Municipal Utility District
    (EBMUD)
East Bay Municipal Utility
    District (EBMUD), 71, 131,
    194
    Briones Reservoir, 133(f)
East of Eden, 24, 73
ecosystem restoration, 84, 218,
    262–66
Edward C. Little Water Recycling
    Facility, 226
Eel River, 59(f), 60, 133
"effluent-dominant" rivers, 83
electric-power generation, 98, 146
El Niño, 27(f), 143
    Southern Oscillation, 24–26
Endangered Species Act (1993), 150,
    157, 207, 211
energy consumption, 145–46
energy use, 146
environmental mitigation, 166
environmental organizations, 213
Environmental Protection Agency
    (EPA), 78, 182, 220
    standards for disinfectants,
        192–93
environmental protection laws, 207
Erin Brockovich (movie), 183
Escherichia coli (fecal coliform)
        bacteria, 195
estuary ecosystem, 169
Evans, Christopher, 191
evaporation, 12, 15, 21, 24, 87, 167,
    169, 171–73, 176, 209, 235, 237,
    241, 264
evapotranspiration, 14(f)
extinction of species, 147–59
ExxonMobil, 227

farmland drainage, 170
Feather River, 62, 91, 95, 100
  tributaries of, 98
  watershed, 108
Federal Energy Regulatory
  Commission (FERC), 246
Feinstein, Dianne, 245
Fenner Valley Water Authority, 244
Fertile Crescent of the Middle
  East, 166
fertilizers, 163
filter membranes, for treatment of
  seawater, 266
fire, 13
Fish and Game Code (1937), 158,
  209
Fish and Game Law (1937), 207
fishes
  anadromous, 43, 57
  artificial spawning, 155(f)
  Chinook Salmon *(O. tshawyt-
    scha)*, 57, 61, 73, 149, 151, 154,
    159
  Coho salmon *(Oncorhynchus
    kisutch)*, 57, 67, 149, 151
  Cui-ui *(Chasmistes cujus)*, 66
  Cutthroat Trout *(O. clarki)*, 58,
    66–67
  Delta Smelt *(Hypomesus
    transpacificus)*, 43, 150, 211,
    216, 218
  endangered species, 154
  Flow Criteria Report for
    protection of, 219
  Golden Trout *(Salmo aguabonita)*,
    77, 77(f)
  Green Sturgeon *(Acipenser
    medirostris)*, 58
  habitat restoration, 71
  hatcheries, 154
  Longfin Smelt *(Spirinichus
    thaleichtys)*, 211

migration of, 153
Owens Pupfish *(Cyprinodon
  radiosus)*, 80
Owens Sucker *(Catostomus
  fumeiventris)*, 80
Pacific Lamprey *(Entosphenus
  tridentatus)*, 44, 149
Sacramento Splittail *(Pogonichthys macrolepidotus)*, 43
salmon, 43
Speckled Dace *(Rhinichthys
  osculus)*, 80
Steelhead Trout, 57, 73, 149, 254
Steelhead Trout *(Oncorhynchus
  mykiss)*, 44, 67
tilapia, 175(f)
Tui Chub *(Gila bicolor)*, 67
fish hatcheries, 108, 154, 156(map)
fishing, recreational, 63
fish ladders, construction of, 154
flood control, 61, 70, 100, 147, 255
flood protection, 33, 220, 251–52, 255
floods, 31–36, 149
  100-year flood, 32, 252
fluoride content, in water system,
  199
fluoride toothpastes, 197–98
fogs, coastal, 14, 55
Folsom Dam, 108, 111, 252
Folsom South Canal, 131
Food and Drug Administration
  (FDA), 195
food chain, 145, 147, 168, 174, 211
fossil fuels, 142, 146, 170, 185, 196
fracking, xvii, 185–88
Frenchman Lake, 98
freshwater, 15, 43–44, 63, 67, 149–50,
  176, 181–82, 226, 228
freshwater marsh, 37, 41
freshwater springs, 79
Friant Dam, 40, 70, 108, 113, 113(f),
  159, 247, 264(f)

Friant–Kern Canal, 113, 113(f), 243
Friant Water Users Authority, 115,
    263
fuel efficiency, 184–85

geological layering, 187
Giardia parasite, 195, 199–201
    in dogs, 202
glaciers, 15, 33
    California, 143
    climate change, affect of, 143
    Sierra Nevada, 16, 143
Globally Important Bird Area,
    77–78
global warming, 33, 142, 145, 172
Gold Country rivers, 182
Golden Trout *(Salmo aguabonita)*, 77,
    77(f)
Golden Trout Creek, 77
Golden Trout Wilderness, 76
Goose Lake, 40, 61
Grand Canyon, 71
"Grassland Area" farms, 171
Gravelly Ford, 113
graywater, reuse of, 227–31
Great Basin, 64
Great Basin desert, 6
Greater Sandhill Crane *(Grus
    canadensis)*, 173
Great Valley, 47
greenhouse effect, 142
greenhouse gases, 142, 146, 196
Grizzly Bears *(Ursus arctos)*, 46
Grizzly Creek Redwoods State
    Park, 58
groundwater, xiii, 16, 50–54, 77, 188
    advantages over reservoirs, 241
    aquifers, 227
    banking of, for future with-
        drawals, 121
    brackish, 266
    change in level of, 240(map)

contamination of, 182
    decontamination of, 184(f)
    Groundwater Management Act
        (2014), 239
    industrial pollutants, 183
    irrigation with, 50, 51(f)
    land subsidence from pumping
        of, 180(map)
    lawsuit against contamination
        of, 183
    management of, xvi, 177, 241
    misconceptions about, 52–54
    overdrafting of, 54
    percolation of, 54, 239
    pollution of, 126, 165, 223
    pumping of, 180(map), 181
    recharge of, 83, 225–26
    remote sensing satellite systems
        for monitoring, 179
    as renewable resource, 52
    restoration plan for contami-
        nated, 182–83
    right to pump, 178
    in San Joaquin Valley, 70
    South Coast groundwater
        basins, 178(map)
    storage of, 241–45
    sustainable, 239–41
    Sustainable Groundwater Act
        (2014), xvi, 177
Groundwater Management Act
    (2014), 239
Groundwater Sustainability
    Agencies (GSAs), 239
Gualala River, 60
Gulf of California, 40, 85
Gulf of Mexico, 119

Haggin, James Ben Ali, 112
Harvey O. Banks Delta Pumping
    Plant, 100
Havasu Lake, 118(f)

hazardous waste products, 198
health impacts, of contaminated
 water, 163
herbivores, 16
Hetch Hetchy Aqueduct, 126–31,
 126–27(map)
Hetch Hetchy Reservoir, 128(f)
Hetch Hetchy Valley, 71, 94,
 128(f), 255
 restoration of, 130, 256
Hodel, Donald, 256
hog wallows, 47
Holder, Charles, 44
Hoover Dam, 118, 283
human demands for water, 172
Human Right to Water Act (2012),
 188, 207, 260
Humboldt Bay, 58, 261, 265
Hundley, Norris, Jr., 177
100-year flood, 32, 252
hydraulic dam, 43
hydraulic fracturing. *See* fracking
hydroelectric power generation,
 100, 103, 172
hydroelectric power plants, 98, 103,
 253
 at Moccasin, 129(f)
hydrofluoric acid, 185
hydrogen bonds, 20, 23
 between water molecules, 21(f)
hydrologic regions, of California,
 54–87
 Central Coast region, 73–74
 Colorado River region, 85–87
 drainage basins, 54
 North Coast region, 55–60
 North Lahontan region, 64–67
 precipitation and runoff, 56(map)
 runoff patterns, 54
 Sacramento River region,
  60–63
 San Francisco Bay region, 67–68

San Joaquin River region,
 68–73
South Lahontan region, 78–82
Tulare Lake region, 74–77
hydropower, 144, 255
 licenses, 246

ibuprofen, 189
ice caps, 15–16
ice cores, 143
ice crystals, 20(f)
Imperial Irrigation District (IID),
 115, 120–21
Imperial Valley, 121, 172–73, 176
industrial emissions, 146
industrial groundwater pollutants,
 183
industrial revolution, 141–42
Inland Empire Utilities Agency,
 261, 266
Integrated Water Management
 (IWM), 272–73
Irrigated Lands Regulatory
 Program (2014), 262
irrigation
 channels, 40, 76
 drip, 160, 161(f), 170
 flood, 162(f)
 with groundwater pumping,
  161(f)
 regulated deficit, 238–39
 water sources used for, 167
Irvine and Scripps Institute of
 Oceanography, 191
Irvine Ranch Water District
 (IRWD), 224
Isabella Dam, 77
Isabella Lake, 77

Joint Powers Water Authorities,
 134
Joshua tree *(Yucca brevifolia)*, 16

Kaweah River, 75
Kern and Pixley National Wildlife
   Refuges, 75
Kern County Water Agency, 242
Kern Lake, 37
Kern River, 75, 77, 112, 242
Kern Water Bank, 242, 243(f)
Kesterson drain, 169
Kesterson National Wildlife
   Refuge, 167
Keswick Dam, 110
keystone species, 151
Kings Canyon National Park, 76
Kings River, 75–76, 159
Klamath Hydroelectric Settlement
   Agreement (2010), 157
Klamath Mountains, 55
Klamath River, 11, 58, 154, 157, 253

Lafayette Aqueduct, 131
lagoons, 41, 164, 165(f), 190(f)
Lagunitas Creek, 67
Lahontan Lake, 64
landform provinces, in California,
   5(map)
land grants, 112
land ownership, 112, 177–78, 206,
   239
landscaping, water used for, 236–37
land subsidence, from groundwater pumping, 180(map)
La Niña, 25–26, 27(f)
   See also El Niño
laws and legislations
   Central Valley Project
      Improvement Act
      (CVPIA), 259
   Clean Water Act (CWA), 166,
      207, 262, 266
   Endangered Species Act (1993),
      150, 157, 207, 211
   Fish and Game Law (1937), 207

Groundwater Management Act
   (2014), 239
Human Right to Water Act
   (2012), 188, 207
Improvement Act (1992), 207
National Parks Act, 130
over Mono Lake, 210
Rainwater Capture Act (2012),
   207, 288
Raker Act (1913), 130
Safe Drinking Water Act (1974),
   207, 285
Sustainable Groundwater Act
   (2014), xvi, 177
Water Commission Act (1914),
   206
Water Conservation Act (2009),
   207, 236, 239
Least Bell's Vireo (*Vireo bellii
   pusillus*), 80
Lee Vining Creek, 90, 123(f)
Lone Pine Creek, 201
Longfin Smelt (*Spirinichus
   thaleichtys*), 211
Los Angeles Aqueduct system,
   xiii, 79, 84, 90, 94, 122(map),
   123–26, 158, 201
   cascade, 124(f)
   pipe in the upper Owens Valley,
   124(f)
Los Angeles River, 83–84, 281
Los Padres National Forest, 73
Lower Klamath Lake, 40
Lux, Charles, 112

Madera Canals, 113
Mad River, 58, 253
Mammoth Creek, 4(f)
Mammoth Lakes, 2, 197
Mammoth Mountain, 70, 89
manures, 164, 261
Marshall, James, 62

marshlands. *See* wetlands
marsh vegetation, 265
mass medication, 197–99
McCloud River, 61, 251
Mead Lake, 118, 121, 172
Mendocino coast, 55
Mendota Pool, 113, 114, 114(f)
Merced River, 34(f), 71–73, 114
mercury, 182
methyl tertiary butyl ether
    (MTBE), 184
Metropolitan Water District of
    Southern California (MWD),
    105, 244, 258
Michael, Jeffrey, 218
microfiltration, 192, 225
migration of species, 153, 159
milk production, 164
Mill Creek, 61
    in Lundy Canyon, 35(f)
Miller, Henry, 112
Millerton Lake, 108
Millerton Reservoir, 248
"mine" salts, 166
Minute 319, 119
Mission Basin Groundwater
    Purification Plant, 270
Modoc National Wildlife Refuge,
    47(f)
Mojave Desert, 6, 7(f), 105, 106(f),
    118, 121, 143, 177, 244
Mojave National Preserve, 244
Mojave River, 78, 82
Mokelumne aqueduct, 131–32,
    132(map)
Mokelumne River, 71, 131, 266
Mono Lake, 16, 30, 78–79, 90, 123,
    125, 158, 207–8, 210, 259, 262–63
    litigation over, 210
    phalaropes at, 209(f)
    tufa towers at, 79(f)
Mono Lake Committee, 210

Monterey Bay, 73
    National Marine Sanctuary, 169
Monterey Shale formation, 185,
    186(map)
Morro Bay, 45
Mount Lassen, 61
Mount Tamalpais, 67
Muir, John, 14, 35, 71, 80, 129–30
Mulholland, William, 84
multicellular organisms, 17–18

Napa River valley, 67
National Aeronautics and Space
    Administration (NASA), 179
National Audubon Society, 210
National Marine Fisheries
    Service, 211
National Parks Act, 130
Natural Resources Defense
    Council (NRDC), 115, 195, 219
natural water cycle, 50, 228
Nevada Falls, 6, 73
New Don Pedro Reservoir, 71, 256
New Melones Dam, 71, 108, 111, 248
nitrates, in drinking water, 163
nitrogen, for plant growth, 163
nitrosamines, 193
North Bay, 133–34
North Coast, 54, 55–60
North Lahontan, 54, 64–67
North-of-the-Delta Offstream
    Storage (NODOS) reservoir,
    249

ocean-atmosphere cycle, 26
off-stream reservoirs, 118, 241,
    250–51, 285
oil and natural gas extraction, 185
Oroville Dam, 62, 98
Oroville Lake, 100
O'Shaughnessy Dam, 71, 126, 128(f),
    130, 256

Owens Lake, 40, 81(f), 89, 125, 176, 262
dust control bubblers, 263(f)
Owens River, 78–80, 125, 262
oxygen gas, 16
ozonation, 194

Pacific Decadal Oscillation (PDO),
26
Pacific Gas & Electric, 183–84
Pacific Institute, Oakland, 232, 236,
238–39
Pacific Lamprey *(Entosphenus
tridentatus)*, 44, 83, 149
Pacific Ocean, 1–2, 40, 52, 67, 82, 90
El Niño Southern Oscillation,
24–26
PacifiCorps, 253
Palo Verde Irrigation District
(PVID), 115
Panoche Water District, 170
paper water, 95, 242, 260
Pardee Dam, 131
Parker Dam, 118, 118(f), 283
pea harvesting, in Central Valley,
69(f)
perchlorate (rocket fuel), 182–83
percolating groundwater, 239
Peripheral Canal referendum
(1982), 217
Perris Lake, 105, 107(f)
photosynthesis, 12–13, 16, 174
Pineapple Express, 32
Pine Flat Dam, 159
Pit River, 61
planetary recycling, 13
Point Loma Wastewater Treatment Plant, 226
Poison Control Center, 198
polar ice, collapse of, 143
pollutant "loadings," 171
population growth, impact on
water supply, 203

Portola Valley, 255
Powell Lake, 172
precipitation in California. *See*
rainfall, in California
Proposition 1 Water Bond (2014),
220–21, 242, 245, 251
Public Trust Doctrine, 206–10, 286
Pulgas Water Temple, 127
Pure Water Project (San Diego),
226
Putah Creek, 133
Putah South Canal, 133
Pyramid Lake, 66, 105, 107

Quantification Settlement
Agreement (2003), 173, 288

rain capturing, 228
cisterns and barrels for, 230(f)
See also water harvesting
rainfall, in California
average annual precipitation,
3(map), 6, 82
patterns of, 2, 4, 8
winter rains, 2
rain forests, 55
rain shadows, 6, 7(f), 8, 64
Rainwater Capture Act (2012), 207,
227, 288
Raker Act (1913), 130, 256
reclaimation of land, 43
"reclaimed" water, 223–24
Red Bluff Diversion Dam, 110,
246(f)
Red-legged Frogs *(Rana aurora
draytonii)*, 74
Redwood National Park, 58
redwoods *(Sequoia sempervirens)*, 55,
57(f)
Regional Water Quality Control
Board, 171, 193
regulated deficit irrigation, 238–39

Reisner, Marc, 119
remote sensing satellite systems, 179
respiration of plants, 13
Restore Hetch Hetchy organization, 255–56
reverse osmosis, 192
  filtering, 225
  membranes, 268
rice fields, in California, 165, 165(f)
Rio de los Americanos. *See* American River
riparian vegetation, 11, 147, 154, 163, 262
River City, 32
river ecosystems, 147
river health, bioassessment of, 266
river rewatering plan, implementation of, 262
Rockwell, Robert, 201
Rocky Mountains, xiii, xiv, 85, 91
Roosevelt, Franklin D., 108, 282
runoff, 8, 11, 31, 43, 59, 65, 113, 127, 144
  agricultural, 163
  Sacramento River, 30
  snowmelt, 141
  stormwater, 166
  Tahoe Lake, 66
Russian River, 60, 133

Sacramento Regional Water Authority, 134
Sacramento River, 30, 32, 40, 54, 59, 61, 67, 90, 100, 108, 110, 132, 151, 167, 189, 211, 217, 225, 250, 252(f)
  acidic mine drainage in, 183(f)
Sacramento–San Joaquin Delta, xvii, 44(f), 100, 150, 170, 193, 204, 207, 211, 215
  pumping restrictions in, 257
Sacramento Splittail *(Pogonichthys macrolepidotus)*, 43

Sacramento Valley, 58, 61, 68, 78, 95, 110, 239, 253–54
  rice fields, 62(f)
  severity of extreme drought in, 28(t)
Safe Drinking Water Act (1974), 207, 285
Salinas River, 73, 74(f)
Salinas Valley, 181
saline land, 37
salmon fish, leaping, 150(f)
"Salmon Forever" group, 265
salmon river, disconnected, 152(map)
Salton Sea, 37, 86, 117, 172–77, 174(f), 259
  as man-made mistake, 176
  salinity issues of, 121, 176
Salton Sink, 40
salt ponds, 265(f)
  restoration of, 264
salt water, 43, 52
San Andreas Fault, 67
San Bernardino Mountains, 82, 105, 249
San Clemente Dam, 254
San Diego Bay, 45, 271
San Diego County Water Authority (SDCWA), 120, 134
San Fernando Valley, 84, 124(f)
San Francisco Bay, 40–41, 42(map), 54–55, 67–68, 251
  safe yield of, 95
  salt ponds, 265(f)
  watersheds, 67
  wetlands and urbanization, 43(map)
San Francisco Examiner cartoon, 215(f)
San Francisco Peninsula, 67, 245
San Francisco Public Utilities Commission (SFPUC), 126, 245, 256

water customers, 130
San Gabriel Mountains, 82–83
San Gabriel River, 45, 82, 84
San Joaquin River, 33(f), 40, 55, 67,
    68–73, 89, 108, 112, 113(f), 114–15,
    159, 171, 247, 263, 264(f)
San Joaquin Valley, 90, 95
    drainage canal, 168(f)
    drainage problems, 167
    groundwater pumping, 181(f)
    National Wildlife Refuge, 69(f)
    severity of extreme drought in,
        28(t)
San Luis Drain, 167, 171
San Luis Obispo Creek, 73
San Luis Reservoir, 100, 107, 111,
    249–50, 285
San Pablo Bay, 41, 68(f)
Santa Ana River, 84, 164
    channelized, 85(f)
    watershed, 225
Santa Clara Valley, 204(f)
Santa Cruz Mountains, 67
Santa Margarita River, 84
Santa Monica Mountains, 84
Santa Susanna Mountains, 83–84
Santa Ynez River, 74
Sarin (nerve gas), 198
"Save Our Water" program (2009),
    236
Scripps Institute of Oceanography,
    172, 192
Seal Beach, 45, 84
sea level, rise in, 143, 145
Searsville Dam, 254–55
seawater, 145
    acidity of, 145
    desalination, 269(f)
    intrusion of, 181, 226, 270
selenium, 168–69
SemiTropic Water Bank, 243
Sequoia National Monument, 76

Sequoia National Park, 8, 75
    snow at, 9(f), 77
sewage treatment plant, xiii
sewage water, discharge of, 265
sewer effluents, 265–66
shale oil reserves, 185
shale rocks, 185
Shasta Dam, 108, 110, 110(f), 246, 251
Shasta Reservoir, 61, 246
Sierra Nevada, 4, 7(f), 91, 147
    Cosumnes River, 48
    Ecosystem Project, 151
    glaciers, 16, 143
    snowpacks, 8, 11, 29, 30(f),
        143–44, 231
    snow survey site in, 31(f)
Silicon Valley, 160, 203–4, 274
silicon vapor, 19
Silverwood Lake, 105, 106(f)
Simi Hills, 84
skeletal fluorosis, 198
Smith River, 48, 57, 286
snowfall, 4, 91
snowpacks, 8, 11, 29, 30(f), 143–44,
    223, 231
SoCalWaterSmart.com, 237
sodium fluoride, 198
soil salinity, 166
Solano County Water Agency, 133
Sonoma County Water Agency, 133
Sonora Pass, 66, 71
sources of water, in California,
    135–40(t)
South Coast, 46, 55, 82–85, 117, 151
    groundwater basins, 178(map)
South Lahontan, 55, 64, 78–82
Southwestern Pond Turtles
    (Clemmys marmorata pallida),
    74, 83
spawning, artificial, 155(f)
spring water, 196
sprinkling, of water, 238(f)

Stanislaus River, 71, 108, 248
    New Melones Dam, 111
State Water Plan (2013), xvii, 223,
    268, 272, 288
State Water Project (SWP), xvii,
    62, 95–107, 211
    allocations, 98(t)
    bond debt, 98
    California Aqueduct, 91, 100,
        101(f), 105, 106(f)
    canals and pipelines, 98
    elevation changes of water
        moving in, 103(f)
    facilities, 96(map)
    hydroelectric power plants, 98
    operating costs, 95
    service areas, 97(map)
State Water Resources Control
    Board (SWRCB), 60, 159, 164,
    193, 210
Stealth Toilet, 232
Steelhead Trout *(Oncorhynchus*
    mykiss), 44, 57–58, 61, 67,
    73–74, 83, 149, 151, 157, 254, 255
Steinbeck, John, 73
Stine, Scott, 31
storm drains, xi, 190, 191(f), 228,
    230(f)
stormwater, 82
    capture of, 227–31
    runoff, 166
"Storm Water Detectives" group,
    266
stormwater management, 228
subterranean lakes, 52
subterranean stream water, 239
Suisun Bay, 41, 67, 167
Suisun Marsh, 41
surface area of lakes, 40
surface tension, 22
surface water, 16, 50, 52, 87, 147,
    239

coordination with groundwater
    supplies, 241
    recycling of, 52
Surfrider Foundation, 191
Susie Lake, 201
sustainable ecosystem, 212, 288
Sustainable Groundwater Act
    (2014), xvi, 177
Sweasey Dam, 253

Table Mountain Lake, 47
Tadpole Shrimp *(Lepidurus*
    packardi), 46
Tahoe Lake, 2, 37, 63–64, 65(f), 86,
    185, 201, 266
    runoff, 66, 66(f)
Tehachapi-Cummings County
    Water District, 218
Tehachapi Mountains, 74
Tehama-Colusa Canal, 110
Temperance Flat Dam, 247–48
Tenaya Lake, 30
Thirsty Dotcom, 203
three-salt lake, 78–79
Tijuana Estuary, 45
tiling, 167
Tioga Pass, 71, 90(f)
toilets
    low-flush, 233(f)
    stealth, 232
Tomales Bay, 55, 67
Total Maximum Daily Load
    (TMDL), 171, 188
Tracy Pumping Plant, 111
TreePeople organization, 228
    Elmer Avenue project, 229(f)
    recommendations for rainwater
        capture, 229
tree rings, 143
trichloroethylene (TCE), 183
trihalomethanes (THMs), 193–94
Trinity Dam, 108

Trinity River, 58, 108
trout fishing, 80
Truckee River, 66, 266
Tulare Lake, 37, 40, 55, 74–77, 171, 179
Tule Elk *(Cervus elaphus nannodes)*, 46
Tule fog, 145
Tule Lake, 40
Tule River, 75, 76
tunnel vision, 153
Tuolumne River, 71, 94, 126, 130, 256
    water delivery system, 256
turbines, 98, 147
Twain, Mark, 64
20×2020. See Water Commission Act (1914)
Tylenol, 189

ultraviolet radiation, 193, 225
underground rivers, 52
Upper Newport Harbor, 45(f)
urban landfills, 41
urban retail water, 236
urban water consumption, 115
U.S. Bureau of Reclamation, 108, 154, 169, 249
    on enlarging Shasta Dam, 251
    "Upper San Joaquin River Basin Storage Investigation" report, 248
U.S. Fish and Wildlife Service, 110, 154, 211
U.S. Geological Survey (USGS), 189
U.S. Public Health Service, 197

Van Duzen River, 58
vegetation in California
    natural, 11
    plants watered by mountain creek, 12(f)

riparian, 11
    water absorption, 12
Ventana Wilderness, 73
Vernal Falls, 34(f), 73
Vernalis Adaptive Management Plan, 263
vernal pool, 16, 46–47, 48(f)

Walker Lake, 66–67
    salinity levels in, 67
Walker River, 30, 64, 66
Warren, Earl, 273
washing machine systems, 231–33
wastewater treatment, 224(f), 225–26
water
    as commodity, 257–60
    as essence of life on Earth, 18
    freezing point, 22
    litigation, 243
    as universal solvent, 20
water absorption, by plants, 12
Water Action Plan, 208, 277
water availability, impact on urban development, xiv
water banking, 241–45
water budgets, 172, 241, 272
Water Commission Act (1914), 206
water conservation, 17(f), 125
    in agricultural use, 238–39
    "best management practices" for, 17
    economic efficiency of, 238
    by living things, 17
Water Conservation Act (2009), 207, 236, 239
water crisis, 210
water cycle, xii, xv, 1–18, 52, 190, 228
    climate change and, 141–46

water cycle *(continued)*
  dynamics of, 143
  wheels within wheels, 11–12, 13(f)
water demand
  for agriculture, 11
  in urban areas, 11
water-dependent habitats, xv, 223
water distribution system, xiii, 130
water efficiency, 146, 220, 275
water filtering, 200(f)
water-for-growth paradigm, 204
water harvesting, 221, 228–29
water imports and exports,
    regional, 92(map)
water landscape of California
  groundwater, 50–54
  hydrologic regions. *See*
    hydrologic regions, of
    California
  pristine waterscape, 37–50
water management, 18, 41, 157, 177,
    250
  best management practices, 17,
    235
  California water law, 206–10
  "county of origin" provisions, 206
  integrated, 272–73
  "watershed of origin" provi-
    sions, 206–7
water marketing, xvi, 257
water meters, installation of, 235
water (H_2O) molecules, 19
  hydrogen bonds, 20, 21(f)
water policy, 222, 276
water pollution, 126, 211
  Chino Basin dairy, 261
  by dairy farmers, 164(f)
water quality, xv, xv, xvii, 129, 250
  degradation of, 265
  and health impacts of contami-
    nated water, 163
  monitoring of, 265

nonpoint sources, impact of, 189
  in Santa Monica Bay, 261
  standards for, 212
  testing of, 270
water recycling program, 219, 220,
    221–27, 268
  costs of, 227
  Pure Water San Diego, 192,
    226–27
water reservoirs, xii, xv, 12, 29
  planetary, 15, 15(f)
water reuse, 221–27
water rights, 206
waterscape, pristine, 37–50,
    38–39(map)
watersheds, xiv, 48
  of Colorado River, xiii, xvi,
    116(map)
  expansion of, 89–95
  of Feather River, 108
  overgrazing of, 163
  protection and restoration of,
    220
  San Francisco Bay region, 67–68
  virtual, 91
water strider, 22(f)
water supply system, 4, 21, 90, 146,
    197
  California Water Plan (1957), 203
  domestic, 199
  fluoride content, 198–99
  impact of population growth on,
    203
  options for, 222(f)
  steps for cleaning of, 192
water table, 50, 181
water trades, 257
water transfers, 95, 220, 257–60, 286
water transport system, xiv,
    93(map)
water treatment plants, xiv, 83, 226
water vapor, 13

evaporated from sea surface, 15
water wastage, 234(f)
water wheels, 223
wells, water supply, 182
West Basin Municipal Water
    District, 223, 226, 270
West Basin Replenishment
    District, 226
wetlands, 44, 46
    in California, 75
    reclamation of, 166
    as wildlife habitat, 46
"wet" water, 95
Whiskeytown National Recreation
    Area, 185
Whiskeytown Reservoir, 108
White Pelicans *(Pelecanus
    erythrorhynchos)*, 67
whitewater rafting, 71, 77
wild and scenic rivers system, 48,
    49(map), 50, 58–59, 62, 251
    American River, 63
    Big Sur River, 73

creation of, 207
Kern River, 77
Kings River, 76
wildlife habitats
    restoration of, 264
    wetlands, 46, 261
Willow Flycatcher *(Empidonax
    traillii)*, 77, 80
Wilson, Pete, 257
Wilson, Woodrow, 130
Winnemem Wintu Indians, 251
Winnemucca Lake, 66
World Health Organization, 197

Yellow-billed Cuckoo *(Coccyzus
    americanus occidentalis)*, 77, 80
Yosemite Falls, 72, 72(f)
Yosemite National Park, 71, 90(f),
    126, 128(f)
    restoration of, 256
Yuba River, 62, 90, 266
Yuma Clapper Rail *(Rallus
    longirostris)*, 173

AUTHOR BIOGRAPHY

David Carle received a bachelor's degree from the University of California at Davis in wildlife and fisheries biology and a master's degree from California State University at Sacramento in recreation and parks administration. He was a ranger for the California State Parks for 27 years. He worked at various sites including the Mendocino Coast, Hearst Castle, the Auburn State Recreation Area, the State Indian Museum in Sacramento, and, from 1982 through 2000, the Mono Lake Tufa State Reserve. He taught biology and natural history courses at Cerro Coso Community College in Mammoth Lakes. David has written several books, among them *Traveling the 38th Parallel: A Water Line around the World, Introduction to Earth, Soil and Land in California, Introduction to Fire in California, Introduction to Air in California* (University of California Press, 2013, 2010, 2008, and 2006), *Water and the California Dream: Choices for the New Millennium* (Sierra Club Books, 2003), *Burning Questions: America's Fight with Nature's Fire* (Praeger, 2002), and *Mono Lake Viewpoint* (Artemisia Press, 1992).